ESSENTIAL
MEDICAL GENETICS

ESSENTIAL MEDICAL GENETICS

J. M. Connor MD BSc MRCP
Wellcome Trust Senior Lecturer, and
Honorary Consultant in Medical Genetics

M. A. Ferguson-Smith FRS FRSE FRCP FRCPATH
Professor of Medical Genetics, and
Director of the West of Scotland Regional Genetics Service

University of Glasgow, and
Duncan Guthrie Institute of Medical Genetics
Yorkhill, Glasgow

BLACKWELL SCIENTIFIC PUBLICATIONS
Oxford · London · Edinburgh
Boston · Palo Alto · Melbourne

© 1984 by
Blackwell Scientific Publications
Editorial offices:
Osney Mead, Oxford, OX2 0EL
8 John Street, London, WC1N 2ES
23 Ainslie Place, Edinburgh, EH3 6AJ
52 Beacon Street, Boston
    Massachusetts 02108, USA
667 Lytton Avenue, Palo Alto
    California 94301, USA
107 Barry Street, Carlton
    Victoria 3053, Australia

First published 1984
Reprinted 1985, 1986

DISTRIBUTORS

USA
    Blackwell Mosby Book Distributors
    11830 Westline Industrial Drive
    St Louis, Missouri 63141

Canada
    The C. V. Mosby Company
    5240 Finch Avenue East,
    Scarborough, Ontario

Australia
    Blackwell Scientific Publications
    (Australia) Pty Ltd
    107 Barry Street, Carlton,
    Victoria 3053

British Library
Cataloguing in Publication Data

Connor, J. M.
    Essential medical genetics.
    1. Medical Genetics
    I. Title   II. Ferguson-Smith, M.A.
    616′.042   RB155

ISBN 0-632-01331-1

Computer typeset by SB Datagraphics
Printed in Great Britain by
Spottiswoode Ballantyne, Printers Ltd

Contents

Appendix

Preface

A genetic component is evident in the aetiology of most human diseases. Hence a knowledge of medical genetics is essential for all practitioners of medicine. Only a few years ago such knowledge had to be self-taught but medical genetics is now included in most medical and dental undergraduate curricula. This book was produced to meet the needs of medical and dental undergraduates for an inexpensive yet comprehensive account of basic principles and clinical applications of modern medical genetics. We also hope that it may be of value to postgraduates who qualified prior to the advent of formal medical genetics tuition.

The text is divided into two sections reflecting our preclinical and clinical lecture courses, developed since their inception in 1963. Over these years the growth in knowledge about the mechanisms of heredity in health and disease has been dramatic and increasingly prevention and treatment of genetic diseases is now possible. This is reflected by the development of medical genetics from a purely academic discipline into a clinical speciality. We have chosen human rather than animal examples of disease to illustrate basic principles, and have sought to emphasise practical ways in which these principles can be applied in medical practice, basing what we teach on our everyday experience in the clinics and laboratories which provide the West of Scotland Regional Genetics Service. We thus include what we regard as essential for medical genetics practice and provide, where necessary, references to specialised texts where the reader may find more extensive coverage of areas of particular interest.

Despite the recent rapid advances, there are still many areas of genetics which are ill-understood. We include these recent advances but also take the less usual step for an undergraduate text of indicating the areas which need further research. Thus we hope to discourage the common undergraduate misconception that nothing is left to discover in this or for that matter any other branch of medicine.

JM Connor
MA Ferguson-Smith

Acknowledgements

We wish to thank many people who have influenced the production of this book. These include:

Victor McKusick with whom we both spent the formative periods of our early training in medical genetics.

Our colleagues at the Duncan Guthrie Institute who have worked with us to establish our clinical and laboratory service and especially Dr Nabeel Affara, Dr David Aitken, Dr Elizabeth Boyd and Dr John Yates for their valuable comments about the manuscript.

We are grateful for permission to reproduce the following figures:

Fig. 3.12: The Editor, *Birth Defects Original Article Series*;
Fig. 3.13: The Editor, *Annales de Génétique*;
Fig. 3.14: Professor Peter Pearson;
Fig. 4.2, 4.3, 4.8, and 5.7: The Editor, *Excerpta Medica*;
Figs. 5.4A and 5.4B: The Editor, *Journal of Medical Genetics*;
Figs. 5.12 and 5.13: The Editor, *Cytogenet. Cell Genet.*;
Figs. 7.3 and 7.8: Dr Douglas Wilcox;
Fig. 8.8: The Editor, Karger;
Fig. 14.5: Blackwell Scientific Publications Ltd;
Fig. 14.12: Mr J Devlin;
Fig. 17.10: Dr Robin Winter;
Figs. 19.1 and 19.2: Professor Nick Wald.

Chapter 1
Human Genetics in Perspective

Human genetics is the scientific study of variation and heredity in man, whereas medical genetics is concerned with the application of these principles to the practice of medicine. Although man has always been aware that individuals differ and that children tend to resemble their parents, the scientific basis for these observations was only discovered during the past 150 years. The clinical application of this knowledge is even more recent, with most progress confined to the past 25 years.

Mendel's contribution

Prior to Mendel parental characteristics were believed to blend in the offspring. Whilst this was acceptable for continuous traits such as height or intelligence, it was clearly difficult to account for the family patterns of discontinuous traits such as haemophilia or albinism.

Gregor Mendel (1822-1884), an Austrian monk, studied single clearly defined pairs of contrasting characters in the offspring of the garden pea. He reached three main conclusions:

1. Inheritance is particulate
Inherited characteristics are determined by pairs of hereditary elements (now called genes).

2. Each pair of genes segregates (Mendel's first law)
The two members of a single pair of genes (the alleles) pass to different gametes during reproduction.

3. The gene pairs show independent assortment (Mendel's second law)
Members of different gene pairs assort to gametes independently of one another.

These two fundamental laws may be summarised: alleles segregate, non-alleles assort.

Although Mendel presented and published his work in 1865 the significance of his discoveries was not realised until the early 1900's when three plant breeders, de Vries, Correns and Tschermak independently rediscovered his findings.

Chromosomal basis of inheritance

In 1839 Schleiden and Schwann established the concept of cells as the fundamental living units. Hereditary transmission through the sperm and egg became known by 1860 and in 1868 Haeckel, noting that the sperm was largely nuclear material, postulated that the nucleus was responsible for heredity. Walther Flemming identified chromosomes within the nucleus in 1877 and in 1903 Sutton and Boveri independently realised that the behaviour of the chromosomes during the production of gametes paralleled the behaviour of Mendel's hereditary units. Thus the chromosomes were discovered to carry the genes. At that time the chromosomes were known to consist of protein and nucleic acid and it was not clear which component was the hereditary material.

Chemical basis of inheritance

Pneumococci are of two genetically distinct strains: rough or non-encapsulated (non-virulent) and smooth or encapsulated (virulent). Griffith in 1928 added heat-killed smooth bacteria to live rough and found that some of the rough pneumococci were transformed to the smooth, virulent type. Avery, McLeod and McCarthy repeated this experiment in 1944 and showed that nucleic acid was the transforming agent. Thus nucleic acid was shown to carry the hereditary information. This stimulated intense interest in the composition of nucleic acids which culminated in Watson and Crick's discovery of the double helical structure for deoxyribonucleic acid (DNA) in 1953.

Human chromosomal disorders

By 1890 it was known that one human chromosome (known as the accessory chromosome) did not always have a partner and in 1905 Wilson and Stevens extended this observation by establishing the pattern of human sex chromosomes. At this time there were believed to be 48 chromosomes in each somatic cell. Tjio and Levan refuted this in 1956 when they showed the normal human chromosome number to be 46. In 1959 the first chromosomal disease in man, trisomy 21, was discovered by Lejeune and colleagues. Other abnormalities were found and by 1970, over 20 different human chromosomal disorders were known. The development of chromosomal banding in 1970 increased the ability to resolve small chromosomal aberrations and so by 1980 more than 50 different chromosome abnormalities were known in addition to many normal variants.

Human single gene traits

The knowledge that certain diseases run in families is not new. In the Jewish Talmud from the 5th Century AD the boys in families affected by haemophilia were excused circumcision. However, the mechanism of inheritance and hence the reasons for the observed familial patterns remained obscure until the 20th century.

In 1902 Sir Archibald Garrod (1858-1936), a London physician, presented his studies on alkaptonuria, a rare condition in which patients have arthritis and urine which darkens on standing. He found 3 of 11 sets of parents of affected patients to be blood relatives and, in collaboration with William Bateson (1861-1926), proposed that this was a Mendelian recessive trait with affected persons homozygous for the underactive gene. This was the first disease to be interpreted as a single gene trait. Garrod also conceived the idea that patients with alkaptonuria really represented one extreme of human biochemical variation and that other less clinically significant variations were to be expected. In 1908 Ottenburg and Epstein presented evidence that blood groups were also inherited as Mendelian traits and in 1911 E.B. Wilson (1856-1939) assigned the gene for colour-blindness to the X chromosome, and so made the first gene assignment in man.

There followed numerous descriptions of distinct human single gene traits and in recent years these have been accurately catalogued by Professor V.A. McKusick at the Johns Hopkins Hospital, USA (Table 1.1). At the present time more than 3500 human single gene traits are known.

The dominant traits tend to concern structural and carrier proteins whilst the recessives like alkaptonuria are often enzyme defects. Pauling in 1949 suspected an abnormal haemoglobin to be

Table 1.1 Human single gene traits

	1966	1968	1971	1975	1978	1982	1984
Autosomal dominant	269 (+568)	344 (+449)	415 (+528)	583 (+635)	736 (+753)	934 (+893)	996 (+959)
Autosomal recessive	237 (+294)	280 (+349)	365 (+418)	466 (+481)	521 (+596)	588 (+710)	599 (+741)
X-linked	68 (+51)	68 (+55)	86 (+64)	93 (+78)	107 (+98)	115 (+128)	118 (+137)
Total	574 (+913)	692 (+853)	866 (+1010)	1142 (+1194)	1364 (+1447)	1637 (+1731)	1713 (+1837)
Grand total	1487	1545	1876	2336	2811	3368	3550

Numbers in parentheses refer to loci not yet fully identified or confirmed.

the cause of sickle cell anaemia and this was confirmed by Ingram in 1956 who found a DNA point mutation which altered the haemoglobin polypeptide sequence. This was the first demonstration in any organism that a mutation in a structural gene could produce an altered aminoacid sequence. In 1959 only two abnormal haemoglobins were known; now the number exceeds 325. In 1951, C and G Cori demonstrated the first enzyme defect in an autosomal recessive condition, glycogen storage disease. By 1959 five enzyme defects were known but the number now exceeds 200. The polypeptide product is, however, still unknown in about 85% of human single gene disorders.

Progress has also been made in the assignment of genes to individual chromosomes. Assignment of genes to the X chromosome is readily made on the basis of the characteristic pattern of inheritance. The first autosomal gene to be assigned was thymidine kinase to chromosome 17 by chromosome segegation in man-mouse hybrids by Weiss and Green in 1967 followed by the Duffy blood group to chromosome 1 by Roger Donahue in 1968. Other techniques have allowed the confirmed assignment of more than 247 autosomal genes which together with the 118 confirmed X chromosomal assignments represent about 10% of the known human single gene traits.

Multifactorial inheritance

Sir Francis Galton (1822-1911), Darwin's half first cousin, studied continuous human characteristics such as intelligence and physique. These traits did not seem to conform to Mendel's laws of inheritance and an intense debate ensued with the supporters of Mendel on the one hand and those of Galton on the other. Finally, a statistician, R.A. Fisher (1890-1962), reconciled the two sides by showing that such inheritance could be explained by multiple pairs of genes each with a small but additive effect. Discontinuous traits with multifactorial inheritance such as congenital malformations were explained by introducing the concept of a threshold effect for the disorder; expression only occurred when the genetic contribution passed the threshold. Many human characteristics are determined in this fashion and usually factors in the environment interact with the genetic background.

Clinical applications

Genetically determined disease is becoming an increasingly important part of ill-health in the community now that infections can be controlled, and now that modern medical and nursing care can save many affected infants who previously would have succumbed shortly after birth.

Table 1.2 Important advances in human genetics

Year	Landmark	Key Figure(s)
1839	Cell theory	Schleiden and Schwann
1859	Theory of evolution	Darwin
1865	Particulate inheritance	Mendel
1877	Chromosomes observed	Flemming
1901	ABO blood groups discovered	Landsteiner
1902	Biochemical variation	Garrod
1903	Chromosomes carry genes	Sutton, Boveri
1908	Inheritance of ABO blood groups	Ottenburg, and Epstein
1910	1st US genetic clinic	Davenport
1911	Linkage in Drosophila	Morgan
1911	1st human gene assignment	Wilson
1927	Mutagenicity of X-rays	Muller
1928	Transfection	Griffith
1940	Concept of polymorphism	Ford
1944	Role of DNA	Avery
1946	Mutagenicity of X-rays	Muller
1946	1st UK genetic clinic	Roberts
1947	Transposable elements	McClintock
1949	Sex chromatin	Barr
1953	DNA structure	Watson and Crick
1956	Aminoacid sequence of HbS	Ingram
1956	46 chromosomes in man	Tjio and Levan
1959	1st chromosomal abn.	Lejeune
1960	Prenatal sexing	Riis and Fuchs
1960	Chromosome analysis on blood	Moorehead
1961	Biochemical screening	Guthrie
1961	X inactivation	Lyon
1961	Genetic code	Nierenberg
1968	1st prenatal chromosomal analysis	Breg and Steel
1968	1st autosomal assignment	Weiss and Green
1970	Prevention of Rhesus isoimmunisation	Clark
1970	Chromosome banding	Caspersson
1972	AFP screening	Brock
1973	HLA disease associations	Terasaki
1978	1st DNA diagnosis	Kan
1979	in vitro fertilization	Edwards and Steptoe
1982	1st product of genetic engineering marketed	

Charles B. Davenport of the Eugenics Record Office in New York State began to give genetic advice as early as 1910. The first British genetic counselling clinic was established in 1946 at Great Ormond Street, London by John Fraser Roberts. Public demand has caused a proliferation of genetic counselling centres so that there are now more than 40 in the UK and more than 450 in the USA.

In 1961 the Guthrie screening test for phenylketonuria and certain other inborn errors of metabolism was introduced. All newborn babies in Britain are screened with this test as early diagnosis and treatment of these conditions can prevent permanent handicap.

Prenatal diagnosis with the option of selective termination of pregnancy provides important reassurance for couples at high risk of serious genetic disorders. This was first attempted in 1968 and is now possible for all chromosomal disorders, more than seventy single gene disorders and some multifactorial conditions.

In vitro manipulation of DNA or genetic engineering is starting to have a clinical impact. So far this has been in the synthesis of gene products such as insulin, growth hormone and interferon but increasingly DNA analysis will be used for prenatal and presymptomatic diagnosis and, in the future, direct treatment of faulty genes may be possible.

Summary

Thus during the past twenty-five years there has been a rapid growth in understanding of genetics of both health and disease. For affected families this has already had an impact with improved genetic counselling and increasingly the potential for both therapy and prevention. For the future one might expect continued progress not only along these lines but also into prenatal and preconceptional screening and prevention. This should lead to a reduction in the incidence of all genetic diseases which will benefit not only the individual families at risk but also society in general.

Chapter 2
Physical Basis of Heredity

In 1944 chromosomal nucleic acid was shown to be the carrier of genetic information. In the succeeding years this principle was shown to hold true for all living organisms and details of the structure and function of nucleic acids have gradually been determined.

NUCLEIC ACID STRUCTURE

Two main types of nucleic acid are recognised: DNA (deoxyribonucleic acid) and RNA (ribonucleic acid). Each nucleic acid macromolecule consists of a sugar-phosphate backbone with projecting nitrogenous bases (Fig. 2.1). In DNA the sugar is deoxyribose whereas in RNA it is ribose. The bases are of two types, purines and pyrimidines. In DNA there are two purine bases adenine (A) and guanine (G) and two pyrimidines thymine (T) and cytosine (C). In RNA uracil (U) replaces thymine. Each unit of base, sugar and phosphate is called a nucleotide.

A molecule of DNA is composed of two nucleotide chains which are coiled around one another to form a double helix (Fig. 2.2). The two chains run in opposite directions and are held together by hydrogen bonds between A in one chain and T in the other or between G and C (Fig. 2.3). This pairing is very specific although rarely erroneous combinations may occur. Since A:T and G:C pairing is obligatory the parallel strands must be complementary to one another. Thus if one strand reads ATGA the complementary strand must read TACT. Hence the ratio of A to T is 1 to 1 and of G to C is likewise 1 to 1. Wide variation exists in the $(A+T)/(G+C)$ ratio. Higher plants and animals tend to have an excess of $(A+T)$ and in man the ratio is 1.4 to 1.

In man the total length of DNA in the haploid set of chromosomes is 3000 million base pairs or 3 million kilobases since each kilobase (kb) equals 1000 base pairs. If stretched out this would have a length of 1.74 metres. The average gene is perhaps two kilobases in size thus for a scale comparison if the total DNA was

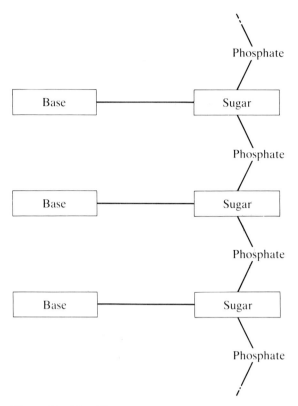

Fig. 2.1 Nucleic acid structure

stretched from Glasgow to London (400 miles) then each gene would occupy about 12 inches.

RNA differs in structure from DNA in several respects:
1. The sugar is ribose rather than deoxyribose
2. Uracil (U) replaces thymine
3. RNA is single stranded
4. Only a single type of DNA is known in man but four types of RNA are found (Table 2.1)

Table 2.1 Types of RNA

Type	Location	Comments
Messenger RNA (mRNA)	nucleus and cytoplasm	variable size, base sequence complementary to transcribed DNA, 1-2% of total RNA
Transfer RNA (tRNA)	cytoplasm	hairpin loop shape, about 40 types, amino acid specific
Ribosomal RNA (rRNA)	ribosomes and nucleoli	about 80% of total cellular RNA
Heterogeneous RNA (HnRNA)	nucleus	high molecular weight, ?mRNA precursor

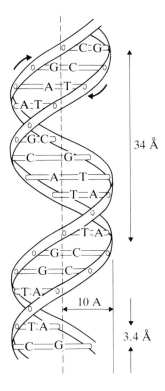

Fig. 2.2 Model of double helix

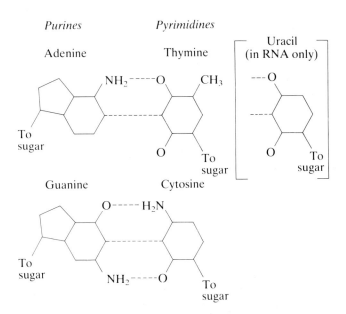

Fig. 2.3 Nucleic acid base pairing

NUCLEIC ACID FUNCTION

Nucleic acids have two major functions: the direction of all protein synthesis and the accurate transmission of this information from one generation to the next.

Proteins, whether structural components, enzymes, carrier molecules, hormones or receptors, are all composed of a series of aminoacids. Twenty aminoacids are known and the sequence of these determines the form and function of the resulting protein. All proteins are encoded in DNA and the unit of DNA which codes for a protein is by definition its gene. Genes vary in size in proportion to their protein products (Table 2.2).

Table 2.2   Examples of genes and their protein products

Protein	Number of aminoacids	Chromosomal location	Approx. gene size (basepairs)
Alpha globin	141	16	850
Beta globin	146	11	1600
Collagen type I (α1 chain)	1000	17	18,000
Collagen (α2 chain)	1000	7	39,000

Each set of three DNA base pairs or triplet codes for an aminoacid. As each base in the triplet may be any of the four types of nucleotide (A,G,C,T) this results in $(4)^3$ or 64 possible combinations or codons. The codons for each aminoacid are given in Table 2.3. Each codon is shown in terms of the messenger RNA and so the corresponding DNA codon will be complementary.

All aminoacids except methionine and tryptophan are coded for by more than one codon: hence the code is said to be degenerate. Three of the 64 codons designate the termination of a message and these are called chain terminators. One codon AUG (methionine) acts as a start signal for protein synthesis With a few possible exceptions this code is identical in all species.

The first stage in protein synthesis is transcription. The two strands of DNA separate in the area of the gene to be transcribed. One strand functions as a template and messenger RNA (mRNA) is formed with a complementary sequence under the influence of the enzyme RNA polymerase (Fig. 2.4). After some processing and modification the mRNA molecule diffuses to the cytoplasm and the DNA strands reassociate.

The next stage of protein synthesis occurs in the cytoplasm and is called translation. Each mRNA molecule becomes attached to one or more ribosomes. As the ribosome moves along the mRNA from the 5′ (five prime) to the 3′ end each codon is recognised by a matching transfer RNA (tRNA) which contributes its aminoacid to the end of a new growing protein chain.

Table 2.3 The genetic code*

First Base	Second Base								Third Base
	U		C		A		G		
U	UUU	phe	UCU	ser	UAU	tyr	UGU	cys	U
	UUC	phe	UCC	ser	UAC	tyr	UGC	cys	C
	UUA	leu	UCA	ser	UAA	stop	UGA	stop	A
	UUG	leu	UCG	ser	UAG	stop	UGG	try	G
C	CUU	leu	CCU	pro	CAU	his	CGU	arg	U
	CUC	leu	CCC	pro	CAC	his	CGC	arg	C
	CUA	leu	CCA	pro	CAA	gln	CGA	arg	A
	CUG	leu	CCG	pro	CAG	gln	CGG	arg	G
A	AUU	ile	ACU	thr	AAU	asn	AGU	ser	U
	AUC	ile	ACC	thr	AAC	asn	AGC	ser	C
	AUA	ile	ACA	thr	AAA	lys	AGA	arg	A
	AUG†	met	ACG	thr	AAG	lys	AGG	arg	G
G	GUU	val	GCU	ala	GAU	asp	GGU	gly	U
	GUC	val	GCC	ala	GAC	asp	GGC	gly	C
	GUA	val	GCA	ala	GAA	glu	GGA	gly	A
	GUG	val	GCG	ala	GAG	glu	GGG	gly	G

Abbreviations for amino acids:

ala	alanine	leu	leucine
arg	arginine	lys	lysine
asn	asparagine	met	methionine
asp	aspartic acid	phe	phenylalanine
cys	cysteine	pro	proline
gln	glutamine	ser	serine
glu	glutamic acid		thr
	threonine		
gly	glycine	try	tryptophan
his	histidine	tyr	tyrosine
ile	isoleucine	val	valine

Other abbreviation:
  stop chain terminators

* Codons are shown in terms of messenger RNA. The corresponding DNA codons are complementary to these. † Start codon for protein synthesis.

The average protein is 300 aminoacids long and these could be coded by 900 base pairs. Genes, however, tend to be larger than predicted (Table 2.2). This excess is due to three factors:
1. Intervening sequences (introns)
2. Post-synthetic modification
3. Regulatory sequences

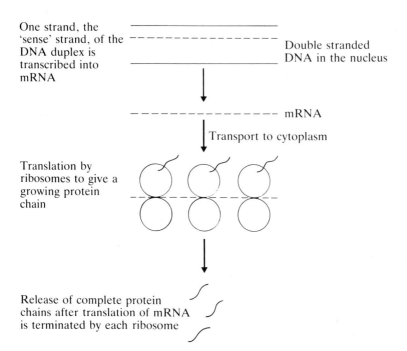

One strand, the 'sense' strand, of the DNA duplex is transcribed into mRNA

Double stranded DNA in the nucleus

Transport to cytoplasm

Translation by ribosomes to give a growing protein chain

mRNA

Release of complete protein chains after translation of mRNA is terminated by each ribosome

Fig. 2.4 Diagram of transcription and translation

The vast majority of genes consist of alternating protein coding segments or exons and non-protein coding segments called intervening seqences or introns (Fig. 2.5). The function of the intervening sequences is unknown. The initial mRNA is a complete transcription of the gene but prior to its entry to the cytoplasm the segments corresponding to the intervening sequences are removed or spliced out. Thus the initial mRNA may be 2-3 times the length of the definitive message. If splicing fails to occcur then an abnormal or no protein product may result. Histone and interferon

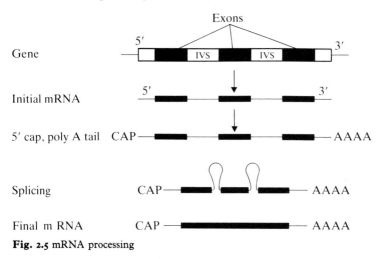

Exons

Gene

Initial mRNA

5' cap, poly A tail   CAP — AAAA

Splicing   CAP — AAAA

Final m RNA   CAP — AAAA

Fig. 2.5 mRNA processing

genes do not contain intervening sequences but these are so far the only known human exceptions. The mRNA 5′ end is blocked or capped and a 3′ poly A tail is usually added to aid transport into the cytoplasm. Many proteins are not in their final form after ribosomal translation. For example, the processed mRNA from the insulin gene produces preproinsulin which is 110 aminoacids in length. The first 24 aminoacids or signal peptide facilitates entry into the endoplasmic reticulum and is then cleaved to produce proinsulin. A sequence of 35 aminoacids (C-peptide, connecting peptide) in the middle of the proinsulin chain is then removed to yield the A chain of 21 aminoacids and the B chain of 30 aminoacids. These chains are joined to one another by disulphide bonds which formed during the initial protein synthesis (Fig. 2.6). Each step in the production of the final protein is important as many proteins are highly dependent for function upon their exact three-dimensional shape which in turn is determined by their aminoacid sequence and post-translational modifications.

In addition to the chain terminator codons, areas of each gene and of neighbouring DNA segments seem to play an important role in regulating transcription and hence synthesis of that protein.

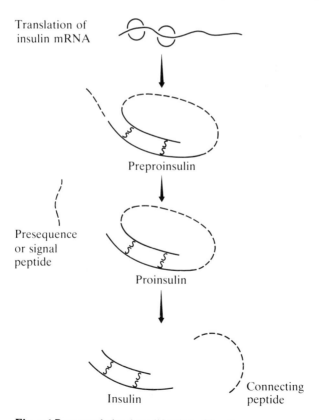

Fig. 2.6 Post-translational modification of insulin

Most estimates indicate 50-100,000 human structural genes. Assuming two kilobases per gene this would only account for at most 200,000 kilobases or 6% of the total (3 million kilobases). Some of the remainder is involved in the regulation of gene expression but most of the rest (perhaps 40% of the total DNA) has no known function and contains large stretches of repetitive DNA. Of this 40% repetitive DNA, 30% is moderately repetitive with several hundred copies (eg. ribosomal DNA) and 10% is highly repetitive with many thousands of copies per genome. One family of highly repetitive DNA, called the Alu family, consists of about 400,000 copies of a 300 base pair sequence interspersed throughout the genome and accounting for about 4% of the total DNA. Satellite DNA is another class of repetitive DNA with numerous tandem repeats. It is found near the centromere of all chromosomes and is especially abundant in chromosomes 1, 9, 16, and the Y chromosome. In contrast, most protein-coding DNA is non-repetitive and has a unique sequence.

DNA REPLICATION

Accurate replication of DNA must occur with each cell division. The two strands separate at a number of points and each strand serves as a template upon which the missing partner can be reconstructed by base pairing with free nucleotides (Fig. 2.7). These nucleotides are bound together by the action of the enzyme DNA polymerase and are hydrogen-bonded to the template strand. Replication proceeds in both directions from each initiation point until the two new strands of DNA are complete. This method of replication is called semi-conservative as one strand of each new DNA molecule has been conserved (Fig. 2.8). This can be shown very effectively by growing cells through two cell divisions in culture medium containing bromodeoxyuridine (BrdU), an analogue of thymine. New growing strands incorporate BrdU, so that at the end of two divisions each chromosome has only one original strand lacking BrdU. This strand can be stained differentially to give the chromosome a harlequin appearance - one dark and one light chromatid (Fig. 2.9). (This technique is put to practical use in the examination of the ability of chemicals to induce sister chromatid exchanges - a form of mutagenicity testing).

MUTATION

Normally replication is completely accurate but errors or mutations can occur. Any such error is copied at subsequent replications

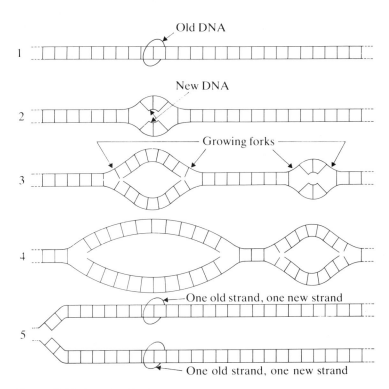

Fig. 2.7 Initiation of replication

unless a reverse mutation occurs. The change in a codon(s) may alter the aminoacid sequence and hence the structure and function of the protein product. Three types of mutation occur:

1. Point substitution
2. Insertion
3. Deletion

In a point mutation a single nucleotide base is replaced by a different nucleotide base. This may lead to no change in the aminoacid coded for by that triplet due to degeneracy of the code (25%) or may result in the substitution of a different aminoacid (Table 2.4). Alteration of an aminoacid codon to a chain terminator can prematurely stop transcription of that gene (5%). The other 70% of point mutations result in variant proteins which may have altered function and/or altered electrophoretic mobility (33%). Insertion or deletion of basepairs generally interferes with transcription by altering the reading frame of the mRNA so that a nonsense message is generated, hence these are called frame-shift mutations. Most mutations are submicroscopic but large deletions or insertions may be visible with the light microscope and are then termed chromosomal aberrations (Chapter 4).

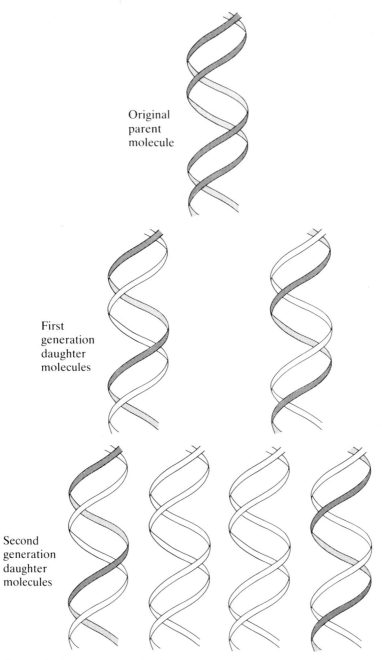

Original
parent
molecule

First
generation
daughter
molecules

Second
generation
daughter
molecules

Fig. 2.8 Semi-conservative replication

Most DNA changes or mutations are spontaneous and unexplained. Certain factors such as mutagenic chemicals and ionizing radiation can increase the rate. In the absence of such agents the mutation rate is in the order of one mutation per 10 million times a gene is replicated.

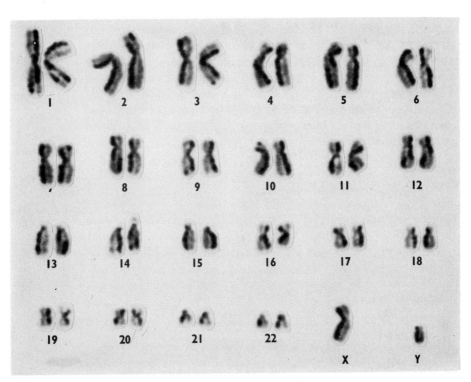

Fig. 2.9 Harlequin chromosomes (Brdu staining)

Table 2.4 Examples of DNA mutation

DNA base sequence	Aminoacid sequence	Comment
CAA TTC CGA CGA	val-lys-ala-ala	Normal sequence
CAA TTT CGA CGA	val-lys-ala-ala	Point mutation with unchanged AA sequence
CAA CTC CGA CGA	val-glu-ala-ala	Point mutation with AA substitution
CAA ATC CGA CGA	val-stop	Point mutation with premature chain termination
CAA —TCC GAC GA	val-arg-leu	Base deletion with frame shift
CAA TTT CCG ACG A	val-lys-gly-cys	Base insertion with frame shift

GENE REGULATION

All nucleated cells have an identical genome yet the relative pattern of gene expression needs to vary widely not only for the differentiation of cells and tissues but also to meet fluctuating demands for the synthesis of different proteins in each cell. Gene regulation in higher organisms is not yet understood. The type of regulatory genes found in bacteria have not so far been identified in man. However, specific triplets occur at the beginning (start) and end (stop) of most mammalian genes. Furthermore, many mammalian genes have coding sequences in their 5′ (front end) flanking regions which are similar to those found in Drosophila and other species. The first of these is AT rich (the Hogness box); the second is the CCAAT box and is found about 70 base pairs to the left of the structural gene. These sequences are believed to be involved in gene regulation.

GENETIC ENGINEERING

Genetic engineering is defined as the artificial formation of new combinations of heritable material.

Three basic components are necessary for genetic engineering:
1. Restriction endonucleases
2. DNA ligases
3. Cloning vectors

1. Restriction endonucleases
Restriction endonucleases are enzymes which cut DNA in a sequence specific manner. They are widespread in microorganisms where they are important in protection against the incorporation of foreign DNA. More than 200 restriction enzymes are now known. Each is named after the organism from which it was first isolated and each will only cleave at a specific DNA sequence - the recognition site (Fig. 2.10). The products of cleavage may have

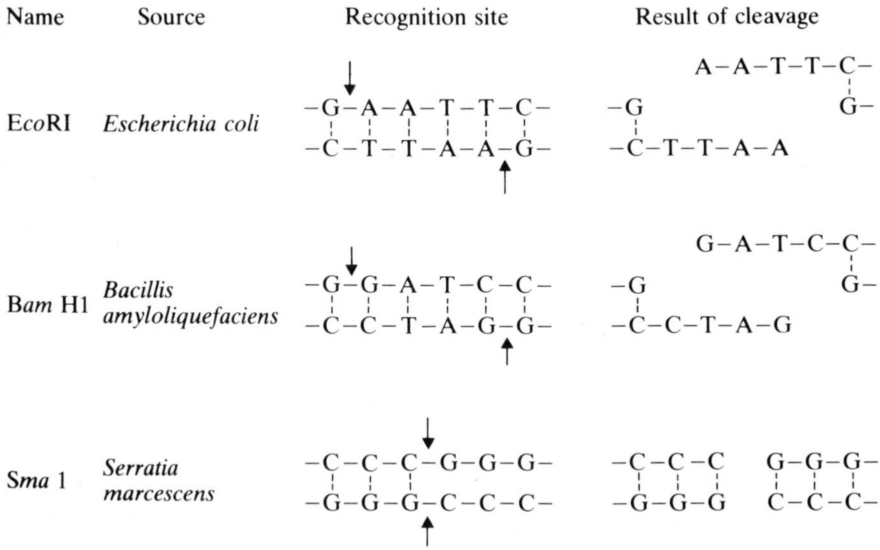

Fig. 2.10 Restriction enzymes

flush (blunt) or staggered (sticky) ends. Thus the enzyme EcoRI will cut DNA at each point where the sequence GAATTC occurs. Simian virus 40 has only one EcoRI recognition site in its single circular chromosome whereas there are about 1000 sites in *Escherichia coli* and about one million sites in man. Digestion of human DNA with EcoRI would thus result in about one million fragments of DNA. These fragments would be of variable length but each would have the same base order at its staggered ends.

Any alteration in the base sequence at a recognition site will result in failure of the enzyme to cut at that site. About 20% of point mutations alter a recognition site usually with loss but sometimes with creation of a new site for a different enzyme. This can sometimes be exploited for the diagnosis of single gene disorders.

2. DNA ligases
DNA ligases are enzymes which join together DNA fragments with either staggered or flush ends.

3. Cloning vectors
Cloning vectors are carrier vehicles for purified DNA sequences and are used to manufacture large numbers of copies of such sequences (amplification). Usually a virus or bacterial plasmid is chosen. The bacterial virus, bacteriophage lambda (λ) is the usual viral vector for cloning. Plasmids are found in all bacteria and are self-replicating circular double-stranded DNA molecules. They are of variable size but contain less than 100 genes. Some of these genes are involved in providing the bacterium with antibiotic resistance. Both the plasmid and viral vectors can replicate autonomously in the host bacterial cell (usually an Escherichia coli) and contain unique sites for a number of restriction enzymes. These unique sites are used as points into which the DNA sequence to be cloned can be inserted.

Recombinant vectors are added to a culture of suitable E. coli which in the presence of calcium chloride readily ingest them. The recombinant DNA is then reproduced in synchrony with the bacterial DNA and large quantities can be produced by growing cultures derived from one cell (amplification). The human DNA can then be isolated from the plasmid DNA by excising the insert and separating the fragments by electrophoresis (Fig. 2.11).

Isolation of a specific DNA sequence

Isolation of a specific human DNA sequence is a prerequisite for DNA diagnosis and therapy. Two approaches may be used to isolate a specific sequence from a complex mixture of DNA sequences:
1. cDNA cloning
2. Cloning of genomic DNA

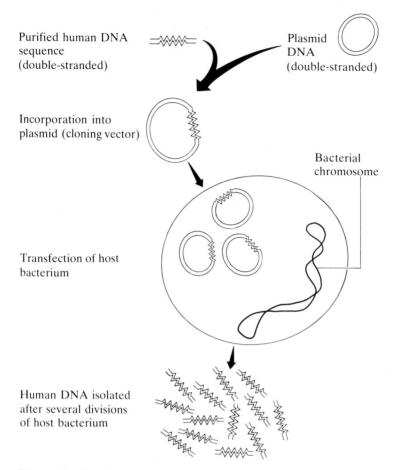

Purified human DNA sequence (double-stranded)

Plasmid DNA (double-stranded)

Incorporation into plasmid (cloning vector)

Bacterial chromosome

Transfection of host bacterium

Human DNA isolated after several divisions of host bacterium

Fig. 2.11 Outline of genetic engineering

1. cDNA cloning

The aim here is to reconstruct the gene of interest from the appropriate messenger RNA. A tissue is chosen which expresses a high level of the gene of interest and only a limited number of other genes. Messenger RNA is isolated from the tissue and the enzyme reverse transcriptase is used to produce single-stranded complementary DNA (cDNA) followed by double-stranded cDNA. This cDNA is then placed into a cloning vector. Since mRNA lacks introns these will also be absent from the cDNA clone as will flanking regulatory sequences. In some circumstances this may have disadvantages and isolation of the genomic sequences may then preferable.

2. Cloning of genomic DNA

Here the starting material is genomic DNA from any convenient nucleated cells. The DNA is cut into fragments with a restriction endonuclease and the fragments are placed into the cut viral vector by sticky end ligation. The recombinant viruses are used to infect bacterial host cells and after amplification these are screened for the appropriate cloned sequence. Cultures of bacteria containing cloned sequences of total genomic DNA are often referred to as genomic libraries. For genomic DNA bacteriophage lambda is used as the vector in preference to a plasmid as it can carry larger pieces of DNA. Genomic DNA cloning can be simplified by fractionating chromosomes carrying the genes of interest using a fluorescence activated cell sorter which separates chromosomes according to DNA content. This method is used to produce chromosome-specific libraries.

If the aminoacid sequence of the gene product is known the artificial assembly of part of the gene may be possible. This oligonucleotide sequence can be used to identify a complementary genomic clone within a chromosomal or whole genomic library.

APPLICATIONS OF GENETIC ENGINEERING TO MEDICAL GENETICS

Current medical applications of genetic engineering relate to diagnosis and therapy.

DNA diagnosis

DNA may be extracted from any nucleated cell. Convenient sources of nucleated cells in man are peripheral blood lymphocytes, cultivated skin fibroblasts, amniotic fluid cells and chorion biopsy material. Once extracted the DNA can be cut into fragments using a restriction endonuclease. The exact number and size of fragments so produced varies from individual to individual in relation to the number of recognition sites for that enzyme. For example in Fig. 2.12 DNA from strand A has 3 recognition sites and so produces 2 fragments whereas B lacks one of these sites and so only produces 1 fragment. This presence or absence of a recognition site is called a restriction fragment length polymorphism (RFLP). The shaded gene in Fig. 2.12 will thus be found in a shorter fragment in DNA strand A than in B.

Fragments of DNA may be separated on the basis of size by agarose gel electrophoresis and individual fragments of interest can be identified by hybridization to a radiolabelled complementary DNA sequence (the probe) (Fig. 2.13). The probe will have been isolated and cloned by one of the techniques described in the previous section.

For some genetic diseases the mutation which causes the disease also causes loss or gain of a recognition site and so can be detectable

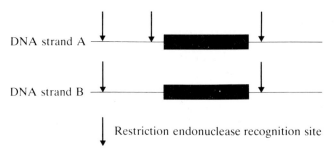

DNA strand A

DNA strand B

Restriction endonuclease recognition site

Fig. 2.12 A restriction fragment length polymorphism

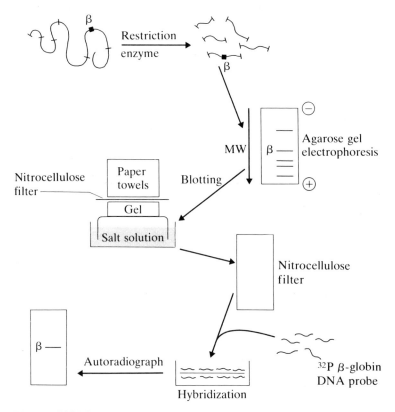

Fig. 2.13 DNA fragment identification (Southern blotting)

directly by changes in the pattern of fragments (e.g. sickle cell disease). Similarly, a large gene deletion would alter the expected fragment size and so is directly detectable (e.g. alpha thalassaemia). Another approach is to identify the disease by the presence of a neighbouring (linked) endonuclease recognition site. All of these approaches are in current use for both prenatal and presymptomatic diagnosis (Chapter 12).

DNA therapy

Genetic engineering is already used to produce proteins of biomedical importance and may in the future be used to replace faulty genes in genetic disorders.

Insulin was one of the first proteins to be so synthesised. Proinsulin cDNA was made from pancreatic mRNA, inserted into a bacterial beta-galactosidase gene carried in a plasmid vector. This was necessary as the cDNA insulin gene lacks regulatory flanking sequences and so the bacterial beta- galactosidase regulatory elements are required to ensure expression of the inserted sequence. Bacteria infected with these plasmids produce large quantities of proinsulin which is purified and converted to pure human insulin by enzymatic removal of the C-peptide.

Table 2.5 Examples of proteins produced by genetic engineering

Insulin
Growth hormone
Somatostatin
Factor IX
Interferon
Vaccines

Table 2.5 indicates proteins which are already in commercial synthesis using genetic engineering.

Chapter 3
Chromosomes

The chromosomes (Gk. chromos = coloured; soma = body) are named from their ability to take up certain stains. They are present in all nucleated cells and contain DNA with its hereditary information. The way in which DNA is arranged with its associated acidic and basic proteins in the chromosome is imperfectly understood. The basic structure is the elementary fibre, 110Å in diameter, composed of repeating units called nucleosomes each made up of 8 histone molecules, around which the DNA molecule is coiled 1 and 3/4 times (Fig. 3.1). The length of DNA joining two nucleosomes is associated with a linker molecule of histone. The elementary fibre of linked nucleosomes is in turn coiled into a fibre of 360 Å diameter. This is the chromatin fibre that can be resolved by examining chromosomes under the electron microscope (Fig. 3.2). The metaphase chromosome has a central scaffold formed of acidic protein to which the chromatin fibre is attached at repeated sequences, so that loops of fibre (Laemli loops - each containing about 200,000 base pairs) radiate out from the scaffold to form the body of the chromosome some 0.6μ in diameter (Figs. 3.3 and 3.4). While the details are unclear, this method of compaction allows DNA to be packaged for cell division. After cell division the chromosome extends and DNA transcription resumes.

Each species has a chromosome complement characteristic in number and form. This is referred to as the species karyotype and the name is also used for a photographic representation of stained chromosomes arranged in order of decreasing length. The act of producing a karyotype is called karyotyping.

KARYOTYPING

In man chromosomes are most conveniently studied in peripheral blood lymphocytes but almost any growing tissue including bone marrow, cultivated skin fibroblasts or amniotic fluid cells can also be used.

Five to ten millilitres of heparinised venous blood is required.

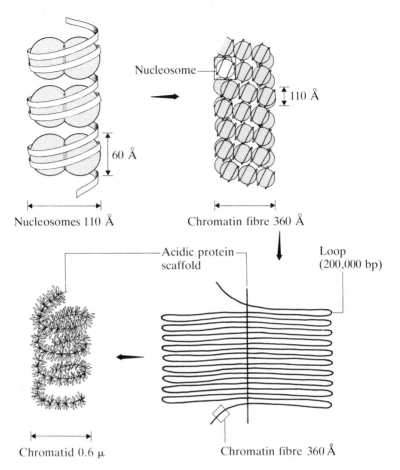

Nucleosomes 110 Å   Chromatin fibre 360 Å

Nucleosome

110 Å

60 Å

Acidic protein scaffold

Loop (200,000 bp)

Chromatid 0.6 μ   Chromatin fibre 360 Å

Fig. 3.1 Nucleosomes
Fig. 3.2 Chromatin fibre
Fig. 3.3 Laemli loops
Fig. 3.4 Chromatid

The heparin prevents coagulation which would interfere with the later separation of the lymphocytes. Samples need to be delivered without delay but a karyotype can still usually be obtained on a blood sample delivered by first class post.

In the cytogenetics laboratory phytohaemagglutinin is added to cultures set up from each blood sample and this stimulates the T lymphocytes to transform and divide. After 48-72 hours incubation, cell division is arrested by the addition of colchicine (which poisons the spindle fibre apparatus), and a hypotonic solution is added to swell the cells and separate the individual chromosomes before fixation. The fixed cell suspension is dropped onto microscopic slides and air dried; this spreads the chromosomes out in one optical plane.

The chromosomes may be stained with numerous stains but for

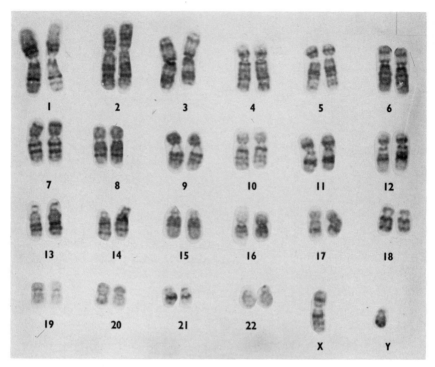

Fig. 3.5 Normal human male karyotype (Giemsa banding, 300 banded)

routine karyotyping G (Giemsa) banding is usually preferred. This produces 300-400 alternating light and dark bands (Fig 3.5) which reflect differential chromosomal condensation. Alternatively, a similar banding pattern can be produced by staining in quinacrine and examining under UV light (Q-banding). Using a method called prometaphase banding it is possible to increase the number of visible bands and thus the resolution (Fig. 3.6). Each chromosome pair has its own characteristic banding pattern. However, the chromosomes can be treated in a number of different ways to show features such as: highly repetitive DNA at the centromeres (C-banding); nucleolus organiser areas with ribosomal mRNA (silver NOR stain); or the late-replicating X chromosome (BrdU incorporation). Some laboratories routinely use reverse (R) banding in which the bands stain in the opposite fashion from that seen with G-banding; this is achieved by heating the chromosomes in a saline buffer.

THE NORMAL HUMAN KARYOTPE

Figure 3.6 shows a normal human female karyotype. In total there are 46 chromosomes which are arranged in order of decreasing size as 23 matching or homologous pairs. They are divided into the

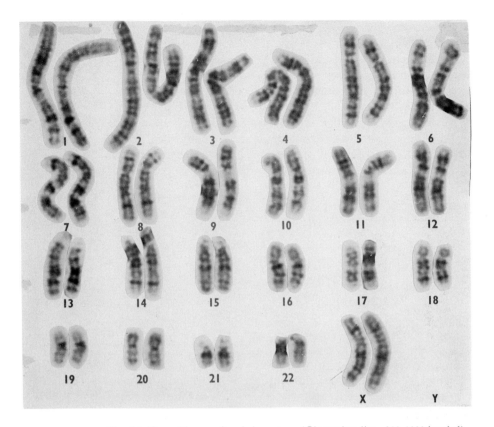

Fig. 3.6 Normal human female karyotype (Giemsa banding, 800-1000 banded)

autosomes (numbers 1 to 22 inclusive) and the sex chromosomes which are two X chromosomes in a normal female. One of each pair of the autosomes and one X is of maternal origin and the other 23 are of paternal origin. In a normal male there are again 46 chromosomes with 22 pairs of autosomes but a different pattern of sex chromosomes namely one X chromosome together with a smaller Y chromosome (Fig. 3.5). One of each pair of autosomes and the X are of maternal origin whilst his father has contributed the Y and the remaining autosomes.

Each chromosome has a narrow waist called the centromere which is the site of attachment of the spindle fibres. The position of the centromere is constant for a given chromosome and three subgroups are identified on the basis of the position of the centromere:

1. metacentric - centromere in the middle
2. acrocentric - centromere close to one end
3. submetacentric - intermediate position of centromere

Each chromosome has a long and a short arm. The short arm is labelled p (from the French word petit) and the long arm as q. The tip of each arm is the telomere.

Chromosomes 1,3,16,19 and 20 are metacentric or nearly so. Chromosomes 13,14,15, 21,22 and the Y are acrocentric and the remainder are submetacentric. With the exception of the Y the human acrocentrics have small masses of chromatin known as satellites attached to their short arms by narrow stalks or secondary constrictions, which mark the location of the ribosomal genes.

Modern banding techniques allow precise identification of any given chromosome. Prior to banding this was not always possible and chromosomes of similar size and shape were grouped into seven groups called A to G (Table 3.1).

Table 3.1 Pre-banding chromosomal groupings

Group	Chromosomes
A	1-3
B	4,5
C	6-12,X
D	13-15
E	16-18
F	19-20
G	21,22,Y

Karyotypes may be described using a shorthand system of symbols (Paris nomenclature). In general this has the order:
1. Total number of chromosomes
2. Sex chromosome constitution
3. Description of abnormality

Thus a normal female karyotype is 46,XX whereas that of a normal male is 46,XY. Table 3.2 lists the other commonly used

Table 3.2 Symbols used for karyotype description

p	short arm
q	long arm
pter	tip of short arm
qter	tip of long arm
del	deletion
der	derivative of a chromosome rearrangement
dup	duplication
i	isochromosome
ins	insertion
inv	inversion
r	ring chromosome
t	translocation
/	mosaicism
+/-	before a chromosome number indicates gain or loss of that whole chromosome
+/-	after a chromosome number indicates gain or loss of part of that chromosome

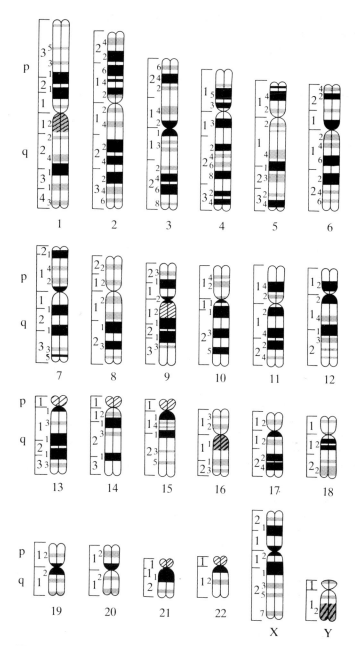

Fig. 3.7 Human idiogram

symbols. A standardised numbering system is used for the bands seen with G-banding (Fig. 3.7). This permits accurate description of chromosomal breakpoints and is useful for describing the location of genes in the chromosomal map.

CHROMOSOME HETEROMORPHISMS

Minor heritable differences in the appearance of the chromosomes are detectable in at least 30% of the population. Such differences are called heteromorphisms and are examples of genetic polymorphisms (Chapter 10). The size polymorphisms usually involve repetitive DNA and the degree of variation shows a normal distribution. With the exception of the fragile site on Xq none of these are associated with clinical problems. Four main groups of chromosome heteromorphisms are known:

1. Size of Yq
2. Size of centromeric heterochromatin
3. Satellite polymorphisms
4. Fragile sites

1. Size of Yq
The commonest polymorphism relates to the length of the long arm of the Y chromosome. About 10% of males have a Y which is obviously longer or shorter than usual (Fig. 3.8). The long arm of the Y contains non-transcribed repetitive DNA and fluoresces intensely under ultraviolet light with dyes such as quinacrine (Q-banding). This fluorescence may be visible in an interphase nucleus and is referred to as the Y-chromatin (Fig. 3.9).

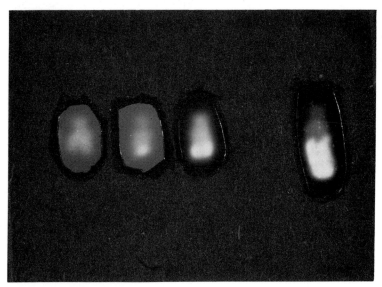

Fig. 3.8 Yq polymorphisms

2. Size of centromeric heterochromatin
Variations in the size of the centromeric heterochromatin are relatively frequent for chromosomes 1,9, and 16. Fig. 3.10 shows a

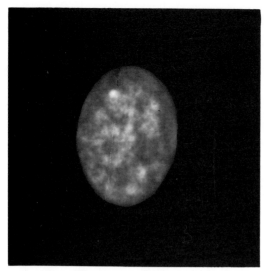

Fig. 3.9 Flourescent Y chromatin

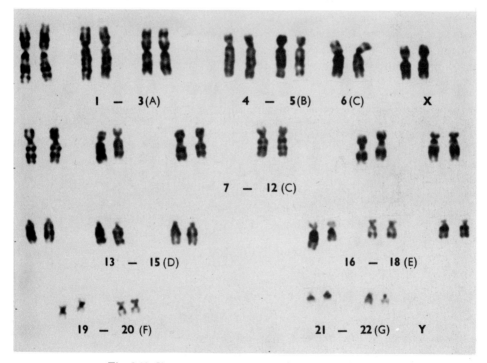

Fig. 3.10 Chromosome 16 centromeric heterochromatin polymorphism (16qh +)

large chromosome 16 centromeric heterochromatin and as seen in Fig. 3.11 this had been inherited by many healthy family members.

3. Satellite poymorphisms
Variation in size of the satellites and in the degree of intensity with

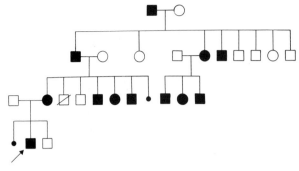

Fig. 3.11 Pedigree of the family with 16qh+ (Fig. 3.10)

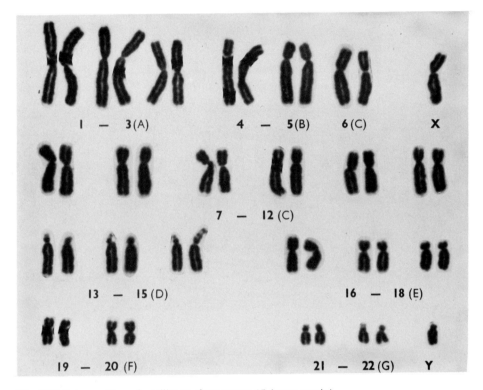

Fig. 3.12 Polymorphism of satellite on chromosome 15 (aceto-orcein)

which they stain by Q-banding may be seen for the acrocentric chromosomes 13,14,15,21 and 22 (Fig. 3.12). Much of the variation is due to repetitive DNA but variation in the number of ribosomal genes also occurs.

4. Fragile sites
Constrictions at sites other than the centromere are sometimes seen and these secondary constrictions may be particularly liable to

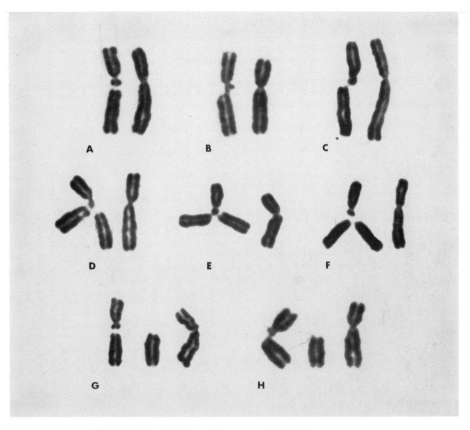

Fig. 3.13 Fragile site on chromosome 2 (2qh')
  A. site shown as gap
  B and C. site shown as chromatid break at gap
  D-F. triradial produced by chromatid breaks in previous division followed by non-disjunction of distal fragment
  G and H. acentric fragments generated by chromatid break

chromatid breaks. Sites include 2q13, 6p23, 9q32, 12q13, 20p11 and Xq27 (Figs. 3.13 and 14.11). These are all induced by antifolate agents in culture. Only Xq27 is associated with clinical abnormality, for it is found to be a marker for a common type of X-linked mental retardation.

CHROMOSOMES IN OTHER SPECIES

The chromosomes appear essentially similar in all races of man. The X chromosome is remarkably constant in size and banding pattern amongst primates. Other chromosomes are more variable and the variation in chromosome number and appearance is in proportion to the timing of evolutionary separation of the species (Table 3.3).

Table 3.3 Chromosomes in different species of animals and plants

Species	Chromosome number
Man	46
Gorilla	48
Mouse	40
Dog	78
Goldfish	94
Maize	10
Garden pea	14
Drosophila	8
E. coli	1

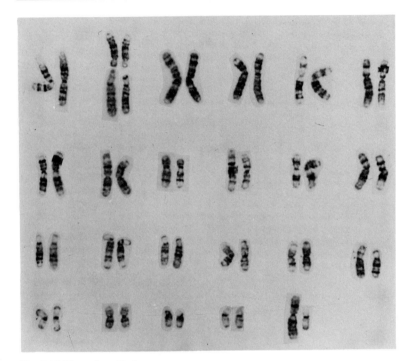

Fig. 3.14 Normal gorilla karyotype

The gorilla, chimpanzee and orangutan have 48 chromosomes and the autosomes are similar to those in man with the exception of human chromosome 2 which appears to have been derived from two ape acrocentrics after separation of the species (Fig. 3.14). Interestingly, the fragile site on human chromosome 2 seems to mark the site of this ancient fusion (Fig. 3.13).

MITOCHONDRIAL CHROMOSOMES

Human mitochondria also have their own chromosomes. A single circular double helix of DNA is found in each mitochondrion.

These are self-replicating and contain genes for transfer RNA, ribosomal RNA and cytochrome C oxidase in their 16,569 base pairs. Human mitochondrial DNA is different from nuclear DNA in respect to the codon recognition pattern for several aminoacids. It is not clear how these differences have evolved.

MITOSIS

Mitosis is a type of cell division by which one cell produces two identical daughter cells. Mitotic cell division occurs in all embryonic tissues and continues at a lower rate in most adult tissues. Thus mitosis is vital for both tissue formation and maintenance. Bacterial cells may undergo mitosis every 20 minutes. In cultured mammalian cells the cycle varies but is usually about 24 hours. Mitosis itself occupies only 20 minutes to 1 hour of the total (Fig. 3.15).

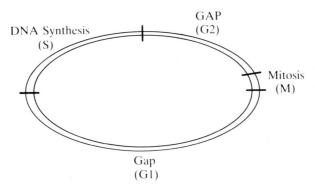

Fig. 3.15 Diagram of cell cycle

Five stages are apparent in mitosis:
1. Interphase
2. Prophase
3. Metaphase
4. Anaphase
5. Telophase

Interphase

A cell which is not actively dividing is in interphase. This phase thus includes Gap 1, S (DNA synthesis) and Gap 2 periods of the cell cycle. During this phase the nuclear material appears relatively homogeneous. Replication of DNA occurs during the S phase so that the nucleus in G2 has twice the diploid amount of DNA present

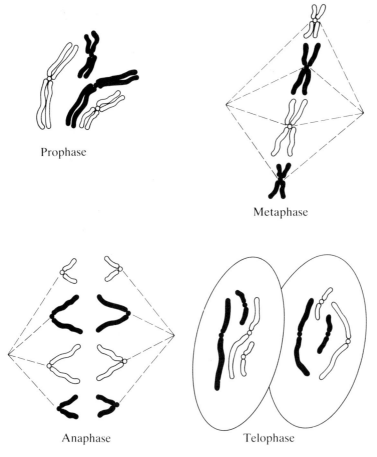

Prophase

Metaphase

Anaphase

Telophase

Fig. 3.16 Diagram of mitosis. Only two chromosome pairs shown; the chromosomes from one parent in outline, from the other in black

in G1. Each chromosome has its own pattern of DNA synthesis and some segments replicate early and some late. The inactive X is always the last chromosome to complete replication. As the cell prepares to divide the chromosomes condense and become visible.

Prophase

Prophase begins when the chromosomes become visible. Each chromosome consists of a pair of long thin parallel strands or sister chromatids which are held together at the centromere. Crossovers with exchange of material may occur at this stage between sister chromatids. Bromodeoxyuridine (BrdU) may be used to demonstrate these sister chromatid exchanges or SCEs (Fig. 3.17). The nuclear membrane disappears. The centriole divides and its two products migrate towards opposite poles of the cell.

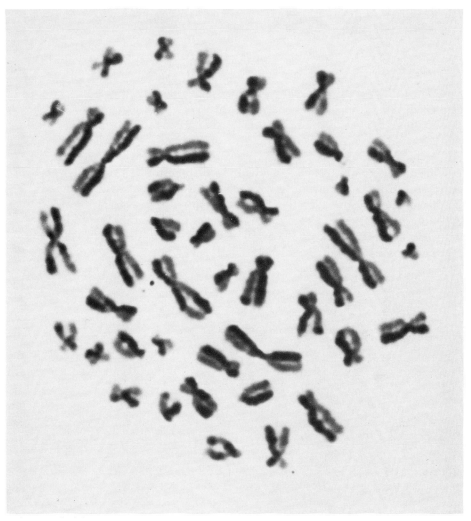

Fig. 3.17 Sister chromatid exchanges

Metaphase

Metaphase begins when the chromosomes have reached their maximal contraction. They move to the equatorial plate of the cell and the spindle forms. The acrocentrics are often clustered at this stage (satellite association). The spindle consists of protein microtubules which connect the centrioles to the centromeres of each chromosome.

Anaphase

Anaphase begins when the centromeres divide and the paired chromatids separate, each to become a daughter chromosome. The spindle fibres contract and draw the daughter chromosomes, centromere first, to the poles of the cell.

Telophase

Telophase starts when the daughter chromosomes reach each pole of the cell. The cytoplasm divides, the cell plate forms and the chromosomes start to unwind. The nuclear membrane reforms at this stage.

Thus mitosis results in two daughter cells each with a identical genetic constitution. Rarely somatic recombination may occur with transfer of segments between homologous chromosomes resulting in homozygosity at gene loci for which the rest of the body cells are heterozygous (Fig. 3.18). This can be an important step in the genesis of some cancers (Chapter 12).

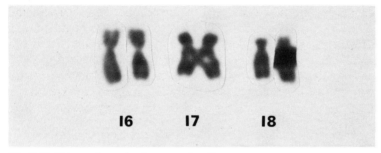

Fig. 3.18 Chiasma formation in a somatic cell

Chapter 4

Gametogenesis

Gametogenesis or the production of gametes occurs in the gonads. The somatic diploid chromosomal complement is halved to the haploid number of a mature gamete in such a way as to ensure that each gamete contains one member of each pair of chromosomes. This reduction is achieved by meiotic cell division. Fusion of the sperm and egg restores the diploid number in the fertilized egg. Meiotic cell division is found only in the gonads and is thus less readily studied than mitosis. Furthermore, as the testis is more accessible than the ovary for biopsy most human information relates to male meiosis. In addition, in the ovary meiosis is arrested in the prophase of first division during fetal life and resumes only after ovulation in the adult.

MEIOSIS

Meiosis consists of two succesive divisions, the first and the second meiotic divisions (Fig. 4.1), in which the DNA replicates only once - before the first division.

First meiotic division (Reduction division)

Prophase of the first meiotic division is complex and five stages can be recognised:
1. Leptotene (threadlike)
2. Zygotene (pairing)
3. Pachytene (thickening)
4. Diplotene (appearing double)
5. Diakinesis (moving apart)

Leptotene starts with the first appearance of the chromosomes. Unlike mitotic chromosomes they consist of alternating thick and thin regions for which each chromosome has its own characteristic pattern (Fig. 4.2). Homologous chromosomes pair during zygotene (synapsis) to form bivalents which are bound closely together by the synaptonemal complex (Fig. 4.3). The X and Y chromosomes

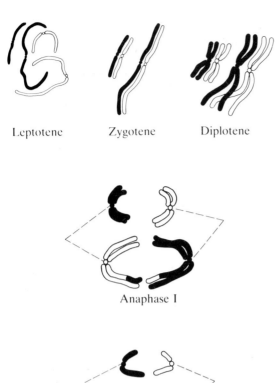

Leptotene   Zygotene   Diplotene

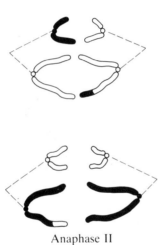

Anaphase I

Anaphase II

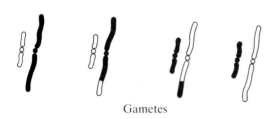

Gametes

Fig. 4.1 Diagram of meiosis. Only two pairs of chromosomes shown; chromosomes from one parent shown in outline, chromosomes from the other in black

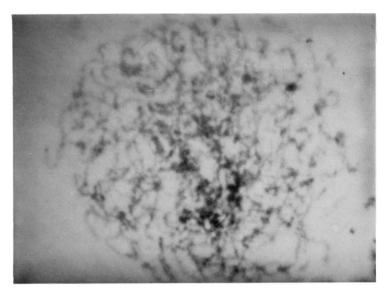

Fig. 4.2 Human primary spermatocyte in leptotene

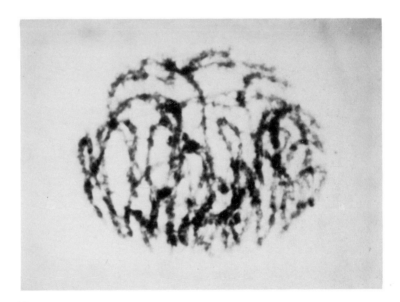

Fig. 4.3 Zygotene

form a sex bivalent which is out of phase with the others and is condensed early as the sex vesicle. This early condensation may be important in preventing crossing over between the X and Y chromosomes. Pachytene is the main stage of chromosomal thickening. The meiotic chromomere pattern corresponds to the banding pattern seen at mitosis (Fig. 4.4). Each chromosome is now seen to consist of two chromatids hence each bivalent is a tetrad of

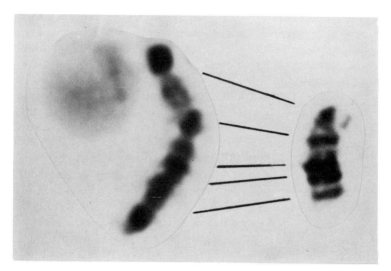

Fig. 4.4 Homology of banding pattern for meiotic (left) and mitotic chromosomes (chromosome 13 shown)

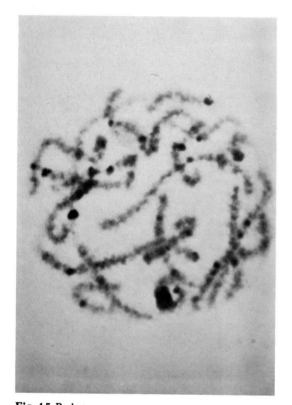

Fig. 4.5 Pachytene

four strands (Fig. 4.5). Satellite association of the acrocentrics occurs at pachytene. Diplotene is very short and difficult to study in man. During diplotene the bivalents start to separate. Although the two chromosomes of each bivalent separate the centromere of each remains intact, so the two chromatids of each chromosome remain together. During longitudinal separation the two members of each bivalent are seen to be in contact at several places called chiasmata (Fig. 4.6). These mark the location of crossovers where chromatids of homologous chromosomes have exchanged material (Fig. 4.7). The chromosomes which have exchanged material are called recombinants in contrast to non-recombinant chromosomes. On average there are about 52 chiasmata per human male cell. At diplotene the sex bivalent opens out and the X and Y can be seen attached to one another by tiny pairing segments at the ends of their short arms suggesting homology of these regions. Diakinesis is the final stage of prophase during which the chromosomes coil more tightly and so stain more deeply.

Metaphase begins when the nuclear membrane disappears and the chromosomes move to the equatorial plane. At anaphase the two

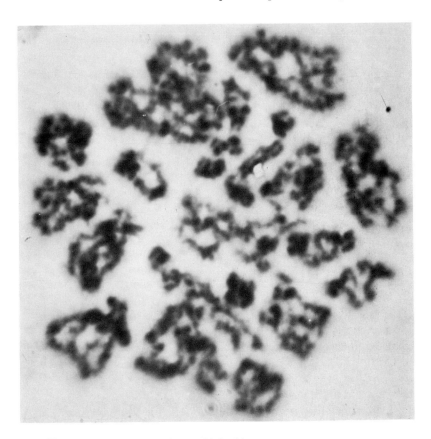

Fig. 4.6 Early diakinesis. Note multiple chiasmata

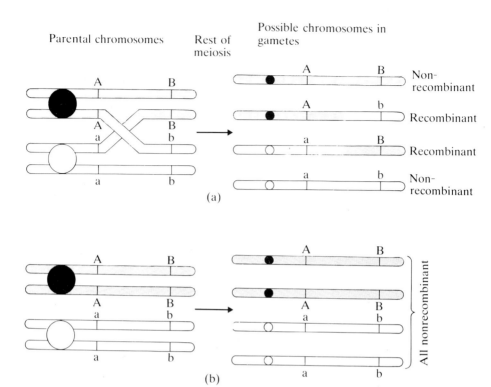

Parental chromosomes   Rest of meiosis   Possible chromosomes in gametes

(a)

(b)

Fig. 4.7 Diagram of crossing-over

members of each bivalent disjoin one going to each pole. These bivalents are assorted independently to each pole. The cytoplasm divides and each cell now has 23 chromosomes each of which is a pair of chromatids differing from one another only as a result of crossing-over.

Second meiotic division

The second meiotic division follows the first without an interphase. It resembles mitosis in that the centromeres now divide and sister chromatids pass to opposite poles. However, the second meiotic division chromosomes are rather more coiled than mitotic ones (relic coiling). The X and Y chromosomes in the male are exceptions and this may be related to the fact that they were not involved in recombination (Fig. 4.8).

Thus meiosis differs from mitosis in several respects as outlined in Table 4.1.

Since the chromosomes assort independently during meiosis this results in 2^{23} or 8,388,608 different possible combinations of chromosomes in the gametes from each parent. Hence there are 2^{46} possible combinations in the zygote. There is still further

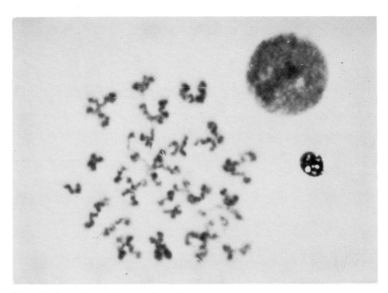

Fig. 4.8 Second meiotic metaphase showing single condensed X

Table 4.1 Comparison of mitosis and meiosis

	Mitosis	Meiosis
Site	all tissues	gonads
Timing	all of life	post puberty in male, suspended until puberty in female
Result	diploid daughter cells	haploid gametes

scope for variation provided by crossing over during meiosis. If there is on average only one crossover per chromosome and a 10% paternal/maternal allele difference then the number of possible zygotes exceeds 6×10^{43}. This number is greater than the number of human beings who have so far existed and so emphasises our genetic uniqueness.

Meiosis has three important consequences:

1. Gametes contain only one representative of each homologous pair of chromosomes.

2. There is random assortment of paternal and maternal homologues.

3. Crossing over ensures uniqueness by further increasing genetic variation.

SPERMATOGENESIS

Spermatogenesis occurs in the seminiferous tubules of the male from the time of sexual maturity onward (Fig. 4.9). At the

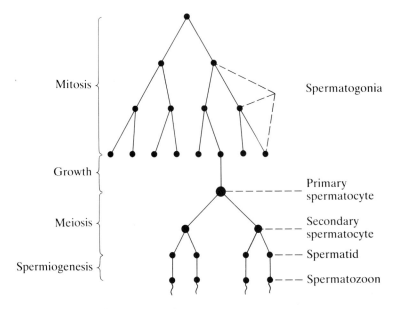

Fig. 4.9 Diagram of spermatogenesis

periphery of the tubule are spermatogonia, some of which are self-renewing stem cells and others which are already committed to sperm formation. The primary spermatocyte is derived from a committed spermatogonium. The primary spermatocyte undergoes the first meiotic division to produce two secondary spermatocytes each with 23 chromosomes. These cells rapidly undergo the second meiotic division each forming two spermatids. The spermatids mature without further division into sperm that are released to the lumen of the tubule. The production of a mature sperm from a committed spermatogonium takes about 75 days.

Normally semen contains 50-100 million sperms per ml. Sperm production continues, albeit at a reduced rate, into old age and the total lifetime production of a male exceeds 10^{12}. The numerous replications increase the chance for mutation and the risk for several single gene mutations has already been shown to be increased in the offspring of older men.

OOGENESIS

In contrast to spermatogenesis the process of oogenesis is largely complete at birth. Oogonia are derived from the primordial germ cells. Each oogonium is the central cell in a developing follicle. By about the third month of fetal life the oogonia have become primary oocytes and some of these have already entered the prophase of first meiosis. The primary oocytes remain in suspended prophase

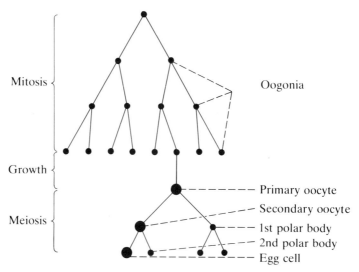

Fig. 4.10 Diagram of oogenesis

(dictyotene) until sexual maturity. Then as each individual follicle matures and releases its oocyte into the Fallopian tube the first meiotic division is completed. Hence completion of the first meiotic division in the female may take over 40 years.

The first meiotic division results in an unequal division of the cytoplasm with the secondary oocyte receiving the great majority in contrast to the first polar body (Fig. 4.10). The second meiotic division is not completed until after fertilization in the Fallopian tube. The second meiotic division results in the mature ovum and a second polar body. The first polar body may also divide at this stage. Thus whereas spermatogenesis produces four viable sperm per meiotic division, oogenesis produces only one ovum.

The maximum number of germ cells in the female fetus is 6.8×10^6 at 5 months. By birth the number is 2×10^6 and by puberty is less than 200,000. Of this number only about 400 will ovulate.

The long resting phase during the first meiotic division may be a factor in the increased risk of failure of homologous chromosomes to separate during meiosis (non-disjunction) in the older mother.

FERTILIZATION

Fertilization usually occurs in the Fallopian tube. As the sperm penetrates the ovum a chemical change occurs which normally prevents the entry of other sperm. After entry the sperm rounds up as a male pronucleus. The ovum now completes the second meiotic division and produces the female pronucleus. These fuse to form the zygote and embryogenesis commences.

Table 4.2 Embryonic and fetal milestones

Stage	Gestation from LMP	Crown-rump length	Comment
Embryo	conception (2 weeks)		
	4 weeks	1mm	Lyonisation, first missed period, chorionic villi develop
	5 weeks	2mm	Neural tube starts to fuse, organ primordia form
	6 weeks	4mm	Neural tube closed, limb buds appear, heart starts to contract, pregnancy test positive, membranes apparent on ultrasound
	8 weeks	3cm	Major organogenesis completed, fetal movements seen on ultrasound, possible chorion biopsy
Fetus	12 weeks	8cm	external genitalia recognisable
	16 weeks	14cm	Optimal time for amniocentesis
	20 weeks	18cm	Optimal time for fetoscopy, fetal movements felt by mother
	28 weeks	24cm	Legal limit for termination
	40 weeks	36cm	Term pregnancy

By a series of mitotic divisions the zygote will produce the estimated 2×10^{12} cells found in the neonate. Table 4.2 summarises the major milestones in embryonic and fetal life of medical genetic importance.

LYONIZATION

Lyonization is the process of inactivation of one member of the pair of X chromosomes in every female cell. It applies to all mammals and in humans probably occurs at or before the 2000 cell stage during the second week of life. Inactivation only occurs in somatic cells since in the germ line both X chromosomes need to remain active. For each somatic cell it is random whether the paternal X or the maternal X is inactivated but the choice is fixed for all subsequent descendents of that cell. As only one X is active in the female the product levels for genes on the X chromosome are similar in females and males where the single X always remains active.

The inactive X is not transcribed with the exception of a small region close to and probably including the pairing region at the tip of the short arm. The inactive X completes its replication later in mitosis than any of the other chromosomes and is thus out of phase with the active X. In females with loss of material from one X chromosome the structurally abnormal X is preferentially inacti-

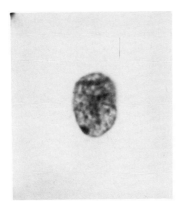

Fig. 4.11 Barr body

vated. In contrast females with an X- autosome translocation preferentially inactivate the normal X, otherwise the inactivation could spread into the autosomal genes leading to autosomal monosomy.

The inactive X remains condensed during most of interphase and is visible in a variable proportion of the nuclei in most tissues as a densely stained mass of chromatin known as the Barr body or X chromatin (Fig. 4.11). Only about 30 % of cells from a buccal smear show X chromatin as this depends upon the stage each cell is at in the cell cycle. If a cell has more than two X chromosomes then the extra ones are also inactivated and more than one Barr body will be seen in each cell. Thus the maximum number of Barr bodies per cell will be one less than the total number of X chromosomes in the karyotype. The sex chromatin may also be seen in 1-10% of female neutrophils as a small drumstick (Fig. 4.12).

Thus a female is a mixture of cells some of which have an active paternal X chromosome and some of which have an active maternal

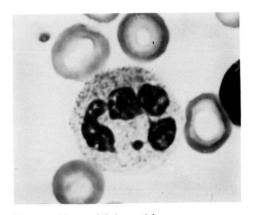

Fig. 4.12 Neutrophil drumstick

X chromosome. The relative proportions vary from female to female due to the randomness of the inactivation process. This accounts for the patchy expression of mutant X-linked genes in carrier females. The tortoiseshell cat is an excellent illustration of the patchy expression of different coat colour alleles carried by the heterozygous female; one X carries the ginger allele, the other the black allele. Examples of the same process are evident in women who are heterozygous for certain X- linked disorders (Chapter 7).

Chapter 5
Chromosome Aberrations

Mutations of the genetic material sometimes involve very large parts of the chromosome and when these are large enough to be visible under the light microscope they are termed chromosome aberrations. With currently available techniques the smallest visible addition or deletion from a chromosome is about 0.2% of the genome or 6 million base pairs. Using the distance from London to Glasgow as the length of the haploid DNA then this would be equivalent to a distance of just over half a mile and on this scale the average gene would be 12 inches in length.

Chromosome aberrations are extremely common as they affect about 7.5% of all conceptions. Most of these are, however, spontaneously miscarried, so the livebirth frequency is 0.6%. Thus about 60% of early spontaneous miscarriages have a chromosomal aberration as do 5% of late spontaneous miscarriages and 4-5% of stillbirths.

Abnormalities of the chromosomes are usually classified into numerical abnormalities, where the somatic cells contain an abnormal number of normal chromosomes, and structural aberrations where the somatic cells contain one or more abnormal chromosomes. They may involve either the sex chromosomes or the autosomes and may occur either as a result of a germ cell mutation in the parent or more remote ancestor, or as a result of somatic mutation in which case only a proportion of cells will be affected.

NUMERICAL ABERRATIONS

Somatic cells contain 46 chromosomes which is the diploid (2n) number. Mature gametes have 23 or the haploid (n) number. A chromosome number which is an exact multiple of the haploid number and exceeds the diploid number is called polyploidy and one which is not an exact multiple is called aneuploidy (Table 5.1).

Table 5.1 Examples of Numerical Chromosomal Aberrations

Karyotype	Comment
92,XXYY	tetraploidy
69,XXY	triploidy
47,XX,+21	trisomy 21
47,XY,+18	trisomy 18
47,XX,+13	trisomy 13
47,XX,+16	trisomy 16
47,XXY	Klinefelter's syndrome
47,XXX	trisomy X
45,X	Turner's syndrome
49,XXXXY	variant of Klinefelter's syndrome

Aneuploidy

Aneuploidy usually arises from failure of paired chromosomes or sister chromatids to disjoin at anaphase (non-disjunction). Alternatively aneuploidy may be due to delayed movement of a chromosome at anaphase (anaphase lag). Thus by either of these mechanisms two cells are produced, one with an extra copy of a

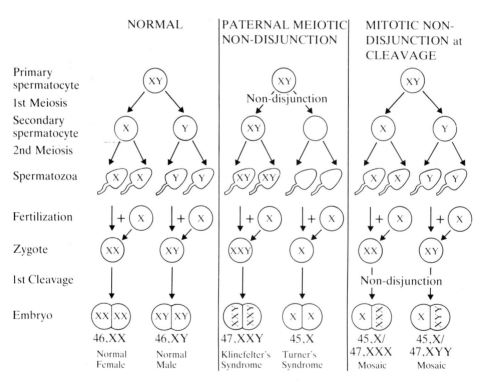

Fig. 5.1 Diagram showing meiotic non-disjunction of sex chromosomes at first meiosis and early cleavage

chromosome (trisomy) and one with a missing copy of that chromosome (monosomy). The cause of meiotic non-disjunction is not known but it occurs at increased frequency with increasing maternal age, with maternal hypothyroidism, and possibly after irradiation or viral infection or as a familial tendency. The cause of mitotic non-disjunction is also unknown and no predisposing factors have been identified.

Aneuploidy can arise either during meiosis or mitosis. Meiotic non-disjunction may occur at either the first or the second meiotic divisions (Fig. 5.1). If the non-disjunction is at the first division then the gamete with the extra chromosome will contain both parental homologues of that chromosome. Whereas if at the second division then both the normal and the extra copy of that chromosome will be either maternal or paternal in origin. Sometimes the origin of the nondisjunctional event can be determined from the knowledge that two alleles at one locus are contributed by one parent, or from the inheritance of chromosomal polymorphisms.

Aneuploidy at a mitotic cell division will result in a mosaic or an individual with cell lines of two or more different chromosomal complements derived from a single zygote.

Polyploidy

A complete extra set of chromosomes will raise the total number to 69 and this is called triploidy (Fig. 5.2). This usually arises from fertilization by two sperm (dispermy) or from failure of one of the maturation divisions of either the egg or the sperm so that a diploid gamete is produced. Thus the chromosomal formula for a triploid fetus (which will usually miscarry) would be 69,XXX; 69,XXY or 69,XYY depending upon the origin of the extra chromosomal set; 69,XXY is the most common.

Tetraploidy or four times the haploid number (4n) is usually due to failure to complete the first zygotic division.

A proportion of polyploid cells occur normally in human bone marrow since megakaryocytes usually have 8-16 times the haploid number. Tetraploid cells are also a normal feature of regenerating liver, and other tissues. They arise by endomitotic reduplication in which the chromosomes divide twice and the cell divides only once.

STRUCTURAL ABERRATIONS

Structural aberrations all result from chromosomal breakage. When a chromosome breaks two unstable sticky ends are produced. Generally repair mechanisms rejoin these two ends without delay. If more than one break has occurred then as the repair mechanisms

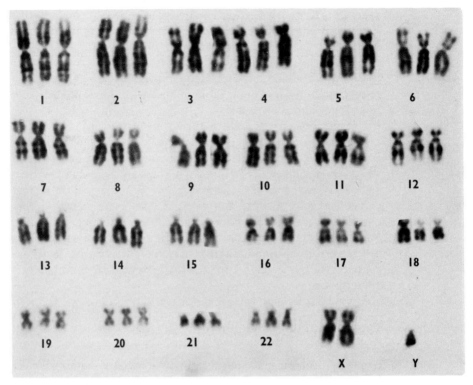

Fig. 5.2 Triploidy

cannot distinguish one sticky end from another there is the possibility of rejoining the wrong ends. The spontaneous rate of chromosomal breakage may be markedly increased by exposure to ionizing radiation or mutagenic chemicals. Chromosomal breakage occurs more often in heterochromatic regions (eg. centromeres and secondary constrictions).

Six types of structural aberration are recognised:
1. Translocation
2. Deletion and ring chromosome
3. Duplication
4. Inversion
5. Isochromosome
6. Centric fragment

Table 5.2 indicates some of the commoner structural chromosomal aberrations.

1. Translocation

A translocation is the transfer of chromosomal material between chromosomes. The process requires breakage of both chromosomes with repair in an abnormal arrangement. This exchange usually results in no loss of DNA and the individual is clinically normal and

Table 5.2 Examples of structural chromosomal aberrations

Karyotype	Comment
46, XY t(5;10)(p13;q25)	balanced reciprocal translocation involving chromosomes 5 and 10 (break points indicated)
45,XX t(13;14)(p11;q11)	centric fusion translocation of chromosomes 13 and 14
46,XY,del(5)(p25)	cri du chat, short arm deletion of 5
46,X,i(Xq)	isochromosome of Xq
46,XX,dup(2)(p13p22)	partial duplication of the short arm of chromosome 2
46,XY,r3(p26 q29)	ring chromosome 3
46,XY,inv(11)(p15q14)	pericentric inversion of chromosome 11

is said to have a balanced translocation. The medical significance is for future generations because a balanced translocation carrier is at risk of producing chromosomally unbalanced offspring.

Three types of translocation are recognised:
a. Reciprocal
b. Centric fusion (Robertsonian)
c. Insertional

a. Reciprocal

In a reciprocal translocation chromosomal material distal to breaks in two chromosomes is exchanged. Either the long or the short arm may break and any pair of chromosomes may be involved (either homologous or non-homologous). Thus in Fig. 5.3 breaks have occurred in the short arm of chromosome 5 and the long arm of

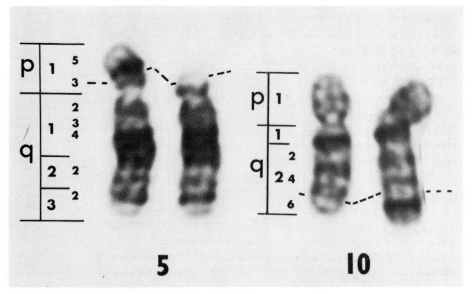

Fig. 5.3 Reciprocal translocation between chromosomes 5 and 10

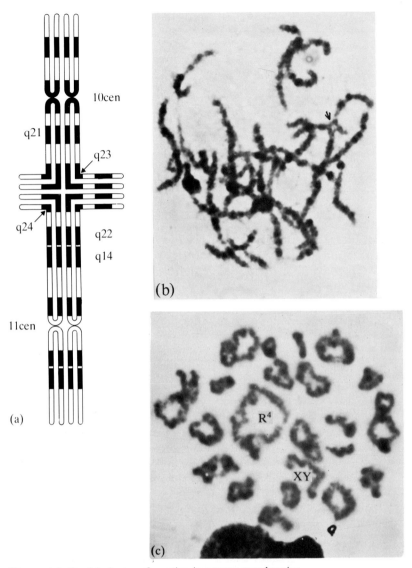

Fig. 5.4 (a). Quadrivalent configuration in a 10;11 translocation

(b). Quadrivalent at pachytene in a 10;11 translocation carrier (arrowed)

(c). Meiotic quadrivalent for a balanced 5;10 reciprocal translocation at diakinesis (R4)

chromosome 10 with reciprocal exchange. The carrier of this balanced translocation is healthy but during gametogenesis unbalanced gametes may be produced. When these chromosomes pair during meiosis a cross-shaped quadrivalent is formed which allows homologous segments to be in contact (Fig. 5.4). This opens into a ring or chain held together by chiasmata. At anaphase these four chromosomes must segregate to the two daughter cells. Twelve possible gametes may be seen. Fig. 5.5 shows the six which result

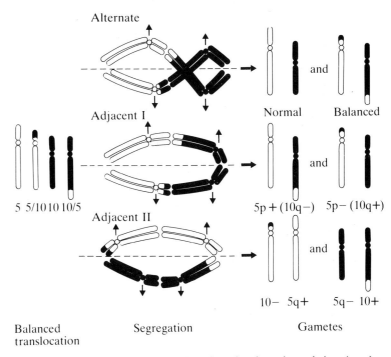

Alternate

Normal | and | Balanced

Adjacent I

5 5/10 10 10/5

5p + (10q−) 5p− (10q+)

Adjacent II

and

10− 5q+   5q− 10+

Gametes

Balanced
translocation

Segregation

Fig. 5.5 Three types of segregation of a balanced reciprocal translation t(5;10) at first meiosis

from a 2 to 2 segregation of these four chromosomes. Of these six possibilities only one gamete is normal and one is a balanced translocation. The other four result in various imbalances of the amounts of chromosomes 5 and 10. Such visible imbalance involves large numbers of genes and affected conceptions may miscarry or if liveborn mental retardation and multiple congenital malformations would be found. Three to one segregation results in a further six gametes but the chromosomal imbalance in each of these is so gross that early spontaneous miscarriage would be invariable. Thus in the liveborn offspring of the carrier of this translocation one would expect a ratio of 1 normal to 1 balanced to 4 unbalanced. In practice some of the unbalanced miscarry, and there may also be selection against the unbalanced gametes so the actual risk of unbalanced offspring is always lower than expected (Chapter 14).

b. Centric fusion (Robertsonian)

Centric fusion translocation arises from breaks at or near the centromere in two acrocentric chromosomes with cross fusion of the products. In most cases the breaks are in the secondary constrictions just above the centromere and so the products are a single chromosome with two centromeres (dicentric) and a fragment with no centromere (acentric) bearing both satellites. An acentric fragment cannot undergo mitosis and will usually be lost at

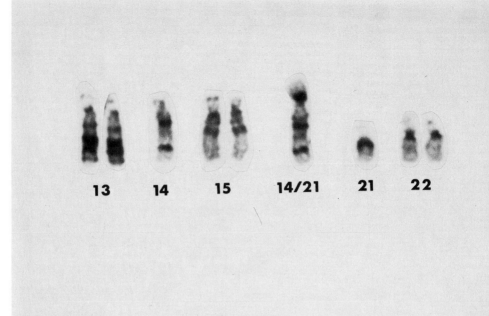

Fig. 5.6 Centric fusion translocation of chromosomes 14 and 21

a subsequent cell division. An alternative possibility is that at least some cases of centric fusion are due to accidental crossovers between homologous sequences on non-homologous chromosomes during first meiosis.

Centric fusion of chromosomes 13 and 14 is the single most frequent type of translocation in man and this is followed in frequency by centric fusion of 14 and 21. Fig. 5.6 shows the karyotype of a balanced 14;21 centric fusion translocation. This combined chromosome is dicentric and the acentric fragment has been lost so leaving only 45 chromosomes in total. Again such an individual is healthy but problems may arise at gametogenesis. When the chromosomes pair during meiosis a trivalent is formed which allows homologous segments to be in contact (Fig. 5.7). At anaphase these three chromosomes must segregate to the gametes. Fig. 5.8 shows the six possible gametes. Only one is normal, one is balanced and four are unbalanced. Again in practice spontaneous abortion and gametic selection result in a lower observed frequency of unbalanced offspring than predicted (Chapter 14).

c. Insertional translocation

For an insertional translocation three breaks are required in two chromosomes. This results in an interstitial deletion of a segment of one chromosome which is inserted into the gap in the other (Fig.

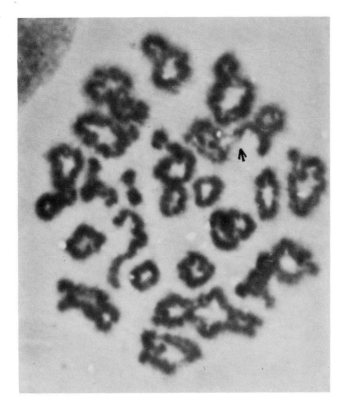

Fig. 5.7 Meiotic trivalent for a t(13;14) centric fusion (arrowed)

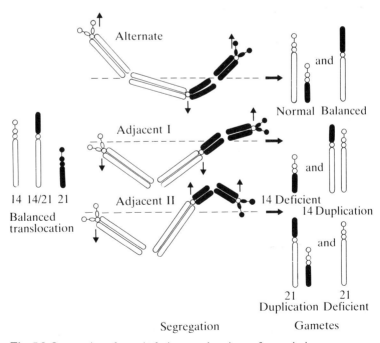

Fig. 5.8 Segregation of centric fusion translocation at first meiosis

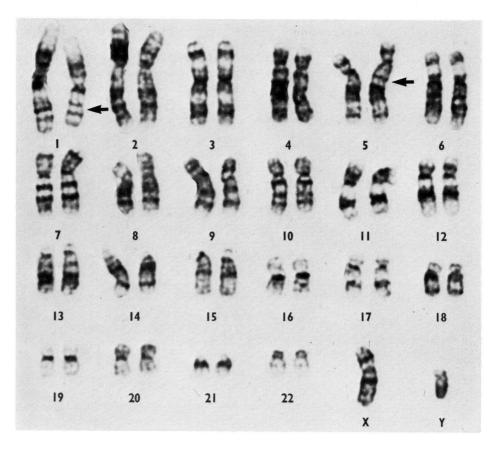

Fig. 5.9 Insertional translocation showing interstitial deletion of band 1q31 and insertion into band 5q13

5.9). Again the balanced carrier is healthy but may produce unbalanced offspring.

2. Deletion and ring chromosome

A loss of any part of a chromosome is a deletion. Deletions usually arise from loss of a portion of the chromosome between two break points (interstitial deletions) or as a result of a translocation. The deleted portion lacks a centromere (an acentric fragment) and will be lost at a subsequent cell division. A ring chromosome arises from breaks in both arms of a chromosome: the terminal ends are lost and the two proximal sticky ends unite to form a ring. If the ring has a centromere then it may be able to pass through cell division.

As the smallest visible loss from a chromosome is about 6000 kilobases individuals with visible deletions are rendered monosomic for large numbers of genes and with autosomal deletions mental retardation and multiple congenital malformations are the rule.

3. Duplication

Duplication is the presence of two copies of a segment of a chromosome. It may originate by unequal crossing-over during meiosis and the reciprocal product is a deletion (Fig.5.10). A duplication can also result from meiotic events in a parent with a translocation, inversion or isochromosome.

Duplications are more common than deletions and are generally less harmful. Indeed tiny duplications at the molecular level (repeats) may play an important role in permitting gene diversification during evolution.

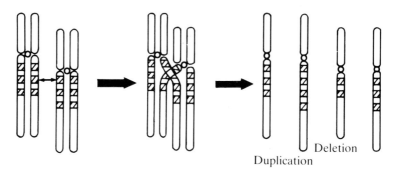

Fig. 5.10 Diagram to show result of unequal crossing over

4. Inversion

Inversions arise from two chromosomal breaks with inversion through 180 degrees of the segment between the breaks. If both breaks are in a single arm then the centromere is not included (paracentric inversion) whereas if breaks are on either side of the centromere it is included (pericentric inversion) (Fig. 5.11). Generally this change in gene order does not produce clinical abnormality. The medical significance lies with the increased risk of generating unbalanced gametes.

Inversions interfere with the pairing of homologous chromosomes during meiosis and crossing over tends to be supressed within the inverted segment. For homologous chromosomes to pair one member must either form a loop in the region of the inversion (Figs. 5.12 and 5.13) or the chromosome arms distal to the inversion fail to pair. For a paracentric inversion, if a crossover does occur within the loop then this will result in a dicentric chromatid and an acentric fragment. Both of these are unstable and fail to undergo cell division and thus do not result in abnormal offspring. In contrast, for a pericentric inversion, if a crossover occurs within the loop then each of the two chromatids produced will have both a deletion and a duplication and abnormal offspring may be produced.

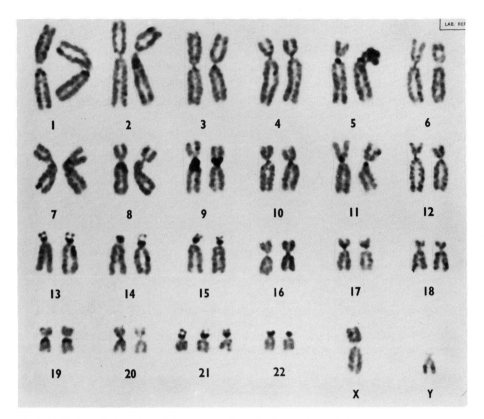

Fig. 5.11 Pericentric inversion of chromosome 9 - 1% of normal population (patient coincidentally has trisomy 21)

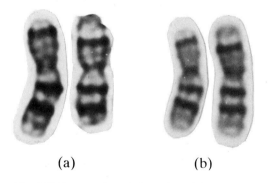

(a)   (b)

Fig. 5.12 Large pericentric inversion of chromosome 7. Normal chromosome in each case to the left.

(a) Parent with a balanced inversion

(b) Abnormal child with duplication (7q32-qter) and deficiency (7p22-pter) resulting from a cross-over within the parental inversion

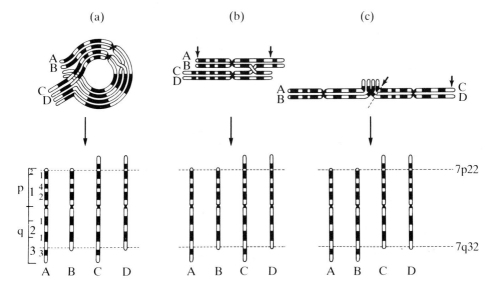

Fig. 5.13 The results of crossing-over at meiosis within (a and b) and outside (c) the pericentric inversion of chromosome 7 shown in Fig. 5.12. A is the normal chromosome 7 and D has the pericentric inversion. In a and b, two types of abnormal recombinant chromosomes are formed (B and C) each with a duplication-deficiency. In c, crossing-over outside the inversion produces no abnormal recombinant

5. Isochromosome

An isochromosome is an abnormal chromosome which has deletion of one arm with duplication of the other. It may arise from transverse division of the centromere during cell division (Fig. 5.14) or from an isochromatid break and fusion above the centromere (in which case it is dicentric). The commonest isochromosome in livebirths is an isochromosome of the long arm of X. This results in clinical abnormality due to short arm monosomy and long arm trisomy. Isochromosomes of Y are also seen in livebirths but for other chromosomes an isochromosome results in an early spontaneous abortion.

6. Centric fragment

Additional small, usually metacentric, fragments are sometimes detected during routine karyotyping. Some are familial and have resulted from a centric fusion translocation arising in meiosis in a parent or ancestor. Provided the centric fragment contains only repetitive and ribosomal DNA, there will be no clinical consequences. Occasionally, transcribed genes are also included in which case there may be associated disability.

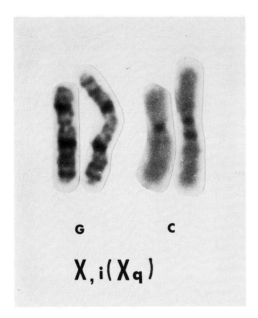

G   C

$X, i(X_q)$

Fig. 5.14 Dicentric isochromosome for the long arm of the X (stained by Giemsa- and C-banding)

OTHER ABERRATIONS

Mosaic

A mosaic is an individual with two or more cell lines which were derived from a single zygote. Thus in an individual with one normal cell line and one cell line with a chromosomal aberration the aberration must have arisen after fertilization. About one percent of patients with trisomy 21 are mosaics with an additional normal cell line. The presence of the normal cell line tends to ameliorate the clinical picture.

If the abnormal cell line is confined to the gonad then an outwardly normal parent may have a high risk of producing abnormal children (gonadal mosaicism).

Chimaera

A chimaera is an individual with two cell lines which were derived from two separate zygotes. This could arise by the early fusion of fraternal twin zygotes or by double fertilization of the egg and a polar body.

Hydatidiform mole

This is an abnormal conception consisting of abnormal chorionic trophoblast in which an embryo is never formed and so the chorionic villi contain no fetal vasculature. The villi become swollen, grape-like structures and the trophoblastic epithelium may undergo malignant change producing choriocarcinoma. Chromosome analysis invariably shows a 46,XX karyotype but the chromosomes are solely derived from the father (androgenetic) and the mole is homozygous at every locus. It seems that the female pronucleus of the fertilised egg degenerates and that the mole develops from diploidisation of the male pronucleus.

Chapter 6
Autosomal Inheritance

With certain notable exceptions (Chapter 12) it is not yet practicable to examine the genetic constitution (genotype) of an individual directly. Generally one infers information about the genotype from the clinical appearance or phenotype. There are several ways of gaining information about the genotype via the phenotype (Table 6.1). Direct examination of the genotype is now possible by DNA sequencing. The molecular tests which infer the genotype from DNA fragment size are therefore included as tests of the phenotype.

Information about families with inherited conditions is stored in the form of a family tree or pedigree. Fig. 6.1 shows some of the

Table 6.1 Ways of examining the phenotype

Clinical examination	
Biochemical tests	— enzyme activity
	— protein electrophoretic mobility
Detection of antigens	
Cytogenetic studies	
Molecular analysis	— linked RFLP
	— direct detection of mutation

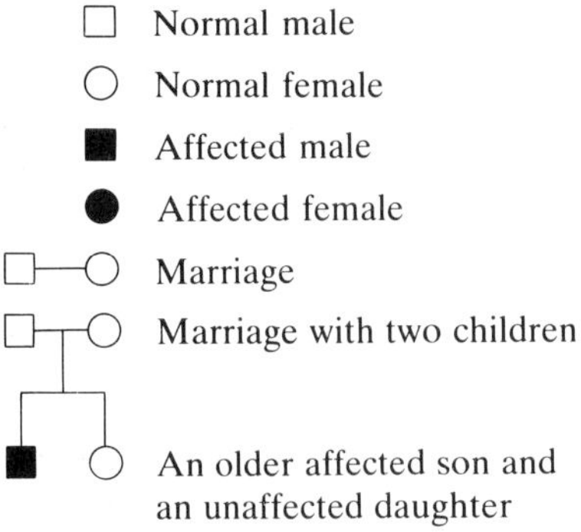

☐ Normal male

◯ Normal female

■ Affected male

● Affected female

☐—◯ Marriage

☐⊤◯ Marriage with two children

■ ◯ An older affected son and an unaffected daughter

Fig. 6.1 Symbols used in pedigree construction (see also Fig. 13.1)

more commonly used symbols and pedigree construction is covered in detail in Chapter 13.

When several members of a family are affected by the same condition then this often but not always indicates that the condition is inherited. There are four main types of inherited disorder chromosomal, autosomal single gene, sex-linked single gene and multifactorial. Chromosomal disorders are covered in Chapter 5, autosomal single gene diseases in this Chapter, sex- linked single gene disorders in Chapter 7 and multifactorial conditions in Chapter 9.

AUTOSOMAL SINGLE GENE INHERITANCE

The 44 autosomes comprise 22 homologous pairs of chromosomes. Within each chromosome the genes have a strict order; each gene occupying a specific location or locus. Thus the autosomal genes are arranged in pairs; one member of maternal and one member of paternal origin. Alternative forms of a gene are called alleles. The normal or common allele is also called the wild type (+). Alleles arise by mutation of the wild type gene and may or may not interfere with the function of that gene. If both members of a gene pair are identical then the individual is homozygous for that locus. If different then the individual is heterozygous for that locus.

Any gene-determined characteristic is a trait. If a trait is expressed in the heterozygote then the trait is dominant whereas if only expressed in the homozygote it is recessive. In some instances the effects of both alleles may be seen in the heterozygote and these are called codominant traits. It is both incorrect and misleading to talk of dominant and recessive genes.

AUTOSOMAL DOMINANT INHERITANCE

This is most easily demonstrated by considering an example. The patient in Fig. 6.2 shows gross facial scarring due to multiple skin cancers. Histologically these are well- differentiated squamous cell epitheliomas but they are unusual by virtue of their numbers and their ability to resolve spontaneously, albeit with scarring, after several months. This patient has a strong family history of similarly affected individuals (Fig. 6.3).

This pedigree shows the typical features of autosomal dominant inheritance. Both males and females are affected in approximately equal numbers. Persons are affected in each generation and males can transmit the condition to males or females and vice versa. Unaffected persons do not transmit the condition. This condition, multiple self-healing squamous epithelioma (MSSE), is due to a

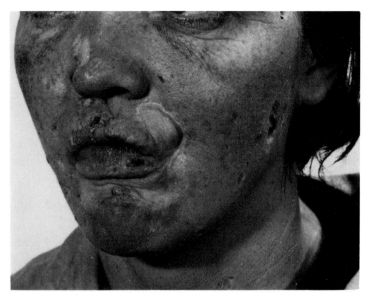

Fig. 6.2 Gross facial scarring from multiple self-healing squamous epithelioma (MSSE)

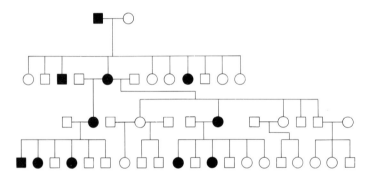

Fig. 6.3 Pedigree of a family with MSSE

single mutant gene. Thus each of the affected persons in this family is a heterozygote and as each has married an unaffected person (normal homozygote) then the expected ratio of affected to unaffected offspring can be seen in Fig. 6.4. It is equally likely for a child to receive the mutant or the normal allele from the affected parent and so on average there is a 1 in 2 or 50% chance that each child of an heterozygous parent will be affected.

In Scotland 11 families with MSSE have been identified. Within these families there are so far 62 affected individuals. Of these 32 are male and 30 female which is close to the expected 1 to 1 sex ratio (chi-square for the difference not significant). The condition never presents before puberty but 90% of males have had their first tumour by 41 years and 90% of females by 34 years, thus one can only be reasonably sure that a person is unaffected after these ages.

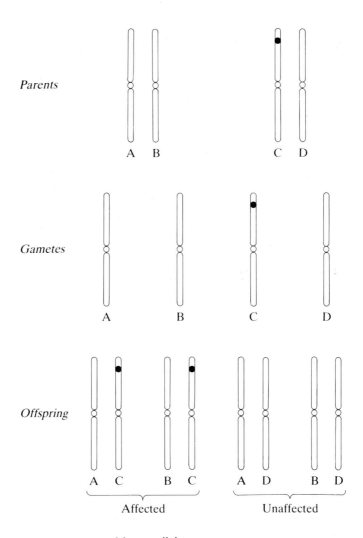

Fig. 6.4 Diagram of autosomal dominant inheritance

The autosomal dominant hypothesis can be tested by comparing the observed segregation ratio with the expected 1 to 1 affected to unaffected. In these 11 families of 74 adults who had a 1 in 2 risk of receiving the mutant allele and who were beyond the age of 90% certainty, 37 were affected and 37 were unaffected. Thus the observed and expected ratios provide further evidence for autosomal dominant inheritance.

Five of these 11 pedigrees are now known to be related by distant ancestors. Furthermore, the few other reported families in the World with this condition all have Scottish ancestry and so there might have been but one original mutation which has now affected upwards of 100 descendents. Although each affected individual has the same mutant gene there is variation in the time of onset and number of tumours. This variable expression is typical of an autosomal dominant trait. Rarely an individual may have the mutant gene yet have a normal phenotype. This is called non-penetrance and is an important exception to the rule that unaffected persons do not transmit an autosomal dominant trait. These individuals can pass on the condition to descendents and so produce a skipped generation. Such non-penetrance is important when counselling for autosomal dominant traits.

So far 1955 autosomal dominant traits are known in man. Some of the commoner and more clinically important of these are shown in Table 6.2. The pedigree pattern in each is similar to MSSE. In

Table 6.2. Autosomal dominant diseases

Dominant otosclerosis	3
Familial hypercholesterolemia	2
Adult polycystic kidneys	0.8
Multiple exostoses	0.5
Huntington's chorea	0.5
Neurofibromatosis	0.4
Myotonic dystrophy	0.2
Congenital spherocytosis	0.2
Polyposis coli	0.1
Dominant blindness	0.1
Dominant congenital deafness	0.1
Others	2.1
TOTAL	10/1000

general they tend to be less severe than recessive traits and whereas recessive traits usually result in defective enzymes, dominants often alter structural or receptor proteins.

Familial hypercholesterolemia is an autosomal dominant trait which affects one in every 500 individuals. Heterozygotes have elevated levels of blood cholesterol and die from premature ischaemic heart disease. This condition is relatively common for an autosomal dominant trait and marriages have occurred between affected heterozygotes. In such matings on average one quarter of the offspring will be unaffected, one half heterozygous affected and one quarter homozygous affected. In the homozygous affected persons the disease is of precocious onset and increased severity with death from myocardial infarction in late childhood. In the few other autosomal dominant traits where homozygous affected individuals have been observed they too have been more severely affected than the heterozygote.

Some autosomal dominant traits are so serious that they usually preclude reproduction (eg. Apert syndrome and progressive myositis ossificans). In this situation neither parent will be affected and the child will be a new mutation. If the child fails to reproduce then the mutant gene goes no further and there will be but one affected individual in the family. For several autosomal dominant traits including Apert syndrome, progressive myositis ossificans, Marfan syndrome and achondroplasia the risk for a new mutation increases with increasing paternal age.

AUTOSOMAL RECESSIVE INHERITANCE

Again this is best demonstrated by an example. The patient in Fig. 6.5 has generalised albinism. This results from a lack of the enzyme tyrosinase which prevents pigment production. The skin is pink red and fails to tan on exposure to ultraviolet light. The hair is dead white and the irides are blue or pink with a prominent red reflex. Visual acuity is generally reduced but otherwise health is unimpaired.

These features may be remembered by recalling the description of Noah's birth from the Book of Enoch:
'She became pregnant and brought forth a child, the flesh of which was white as snow; and red as a rose; the hair of whose head was white, like wool and long; and whose eyes were beautiful. When he opened them he illuminated all the house like the sun.'

The pedigree of the patient in Fig. 6.5 is shown in Fig. 6.6. The parents are both clinically normal yet are heterozygotes (carriers) for the mutant tyrosinase allele. Their other normal allele produces sufficient tyrosinase to permit normal pigmentation. No other

Fig. 6.5 Child with generalised tyrosinase negative albinism. Heterozygous parents have normal pigmentation

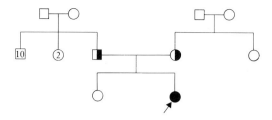

Fig. 6.6 Pedigree of the family shown in Fig. 6.5

individuals are affected in the family. The clinically normal sister may be a heterozygote or be homozygous for the normal allele. Current tests cannot distinguish these two possibilities.

Fig. 6.7 shows the possible offspring for parents who are both carriers for albinism. On average one quarter of their children will be homozygous normal, one half heterozygous and one quarter homozygous affected.

As for MSSE the observed segregation ratio can be compared with the predicted. Two points must be borne in mind when using this approach for a suspected autosomal recessive trait. First, it is unlikely that any single family will have produced sufficient children to give the ratio directly. Second, there is an automatic bias since families only come to medical attention by virtue of an affected child and those carrier parents who by chance produce only unaffected children will be missed. Thus in the 30 sibships with albinism seen in Fig. 6.8, 70 children out of a total of 204 (35%) are affected. A correction for the bias can be made by not counting the first affected child in each family and with this technique the segregation ratio falls to 40 out of 174 (23%) and the chi-square shows that this difference is not statistically significant from the expected 25% (Appendix 1).

For an albino parent each child must receive a mutant allele but if married to a homozygous normal then only unaffected heterozygotes will be produced. If by chance an albino marries a heterozygote then there will be a 1 in 2 chance that each child will be affected. If an albino marries another albino of the same type (i.e. allelic) then only albino children can be produced; albinos of different types produce doubly heterozyous unaffected offspring (i.e. non-allelic).

The majority of parents of children with albinism are not blood relatives (consanguineous) but if they are there is an increased risk for this and other autosomal recessive disorders. Again Noah serves to emphasise this point for his parents Lamech and Betenos were first cousins. The increased risk in this situation is caused by the parents sharing one set of grandparents and the chance that each has inherited the same albino gene from one grandparent (Table 6.3). The proportion of shared genes (coefficient of relationship, r) decreases by one half for each step apart on the pedigree. In highly

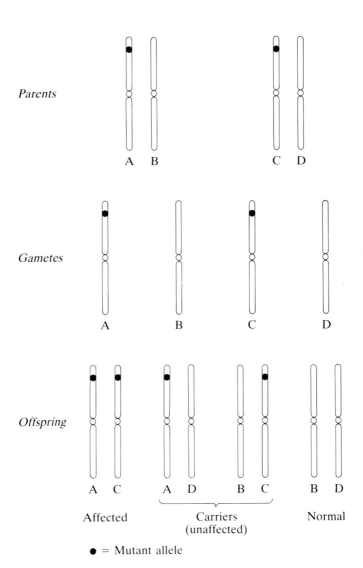

Parents

A B   C D

Gametes

A   B   C   D

Offspring

A C   A D   B C   B D

Affected   Carriers   Normal
            (unaffected)

● = Mutant allele

Fig. 6.7 Diagram of autosomal recessive inheritance

Table 6.3

Degree of relationship	Examples	Proportion of genes in common (r)
1st	parents to child, sib to sib	1/2
2nd	uncles or aunts to nephews or nieces, grandparents to grandchildren	1/4
3rd	first cousins, great grandparents to great grandchildren	1/8

Autosomal Inheritance 73

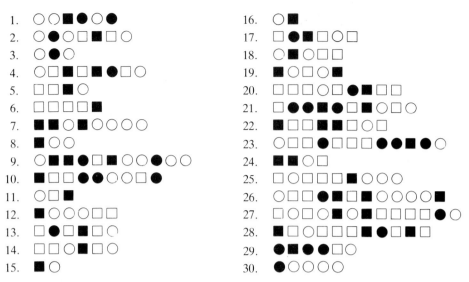

Fig. 6.8 Examples of sibships with generalised albinism

inbred populations the affected person has a susbstantial risk of mating with a carrier. This may result in a pedigree with apparent vertical transmission of an autosomal recessive trait (quasidominant inheritance). Hence parental consanguinity, whilst not a prerequisite, is an important clue that a condition in their child is an autosomal recessive trait.

Generalized albinism due to defective tyrosinase affects one in every 40,000 Caucasian livebirths and the carrier frequency in this ethnic group is 1 in 100. The gene frequency is much higher in certain groups, for example the birth frequency of albinism in the Cuna indians of Panama is 1 in 143 with a carrier frequency of 1 in 6. Thus the ethnic origin of a patient may be an important clue to an autosomal recessive disorder (Table 6.4).

So far 1340 autosomal recessive traits are known in man. Some of the commoner and more clinically important of these are shown in Table 6.5. In about 15% of autosomal recessive traits an enzyme

Table 6.4 Ethnic associations with autosomal recessive diseases

Disease	Ethnic group(s)
Beta thalassaemia	Cypriots, Greeks, Italians, Thais, Indians, Chinese, Turkish, US Negroes
Sickle cell disease	African negroes, Arabs, West Indians
Tay-Sachs	Ashkenazi Jews
Gaucher disease	Ashkenazi Jews
Adrenogenital syndrome	Eskimos
Severe combined immunodeficiency	Apache Indians
Cystic fibrosis	Caucasians

Table 6.5 Autosomal Recessive Diseases

Disease	Frequency/1000 births
Cystic fibrosis	0.5
Recessive mental retardation	0.5
Congenital deafness	0.2
Phenylketonuria	0.1
Spinal muscular atrophy	0.1
Recessive blindness	0.1
Adrenogenital syndrome	0.1
Mucopolysaccharidoses	0.1
Others	0.3
TOTAL	2/1000

defect has already been demonstrated and is to be expected in the remainder. Thus these conditions are inborn errors of metabolism. For many traits not just one but multiple different mutant alleles may occur at the locus (*multiple allelism*). Some, but not all of these, result in sufficient reduction of enzyme activity to produce disease in the homozygous state. An individual who has two different mutant alleles is termed a genetic compound.

Many conditions which were believed to be single genetic entities are now known to be *genetically heterogeneous* (i.e. to have several different genetic causes). For example, in addition to the usual tyrosinase negative form of albinism described in this Chapter, tyrosinase positive and other rare types are known. In some conditions the genetic mimics may have different modes of inheritance and thus pose problems during genetic counselling.

AUTOSOMAL CODOMINANT

The inheritance of the ABO blood groups serves to illustrate autosomal codominant inheritance. For each individual the ABO blood group is determined by a gene locus on chromosome 9. Three main alleles occur A, B and O and so six possible genotypes are found (Table 6.6). Testing with antisera only distinguishes four

Table 6.6 ABO blood groups

Genotype	Phenotype
AA	A
AO	
AB	AB
BB	B
BO	
OO	O

phenotypes A, B, AB, and O. Thus individuals with the genotype AA and AO both give the A phenotype and individuals with the genotype BB and BO both give the B phenotype. Hence the traits for alleles A and B are both dominant to that determined by the O allele, however, individuals with alleles A and B express both.

The pedigree pattern of human codominant traits resembles autosomal dominant inheritance except that both alleles can be distinguished. Some of the commoner and more clinically important traits of this type are listed in Table 6.7. For many of these the

Table 6.7 Autosomal Codominant Traits

Blood groups — ABO, Duffy, Kell Kidd, MNS, Rhesus
Red cell enzymes — Acid phosphatase, adenylate kinase
Serum proteins — Haptoglobin
Cell surface antigens — Human Leucocyte Antigen system (HLA)
Restriction fragment length polymorphisms (RFLPs)

rarest allele at the locus occurs with a frequency of more than 2% and so these are examples of genetic polymorphisms and as such are useful for linkage analysis (Chapters 8 and 10).

SUMMARY OF AUTOSOMAL INHERITANCE

Table 6.8 summarises the principle distinguishing features of autosomal recessive and dominant inheritance.

Table 6.8 Comparison of Autosomal Dominant and Recessive modes of inheritance

Autosomal dominant	Autosomal recessive
Disease expressed in heterozygote	Disease expressed in homozygote
On average 1/2 of offspring affected	Low risk to offspring
Equal frequency and severity in each sex	Equal frequency and severity in each sex
Paternal age effect for new mutations	
Variable expressivity	Constant expressivity in a family
Vertical pedigree pattern	Horizontal pedigree pattern
	Importance of consanguinity

Chapter 7
Sex-Linked Inheritance

The female has two X chromosomes: one of paternal and one of maternal origin. However, with the exception of some genes near the tip of the short arm, one of these X chromosomes is inactivated in each somatic cell (Lyonisation, Chapter 4). This mechanism ensures that the amount of X-linked gene products produced in somatic cells of the female is equivalent to the amount produced in male cells. Dosage compensation applies to all X-linked genes except those near the pairing segment at the end of the short arm such as the steroid sulphatase gene and the Xg red cell antigen gene; females produce nearly twice as much of these gene products as males. In the process of inactivation the choice between the maternal and paternal X homologues is random although once established the same homologue is inactivated in each daughter cell. Thus the female is really a mosaic with a percentage of cells having the paternal X active and the maternal X active in the remainder. Each son or daughter receives one or other X chromosome from their mother.

In contrast the male has only one X chromosome and hence only one copy of each X-linked gene (hemizygous). The Y chromosome probably contains homologous loci for those on the tip of the short arm of the X and also contains important male determinants. In the male the X chromosome remains active in every cell and so any mutant X alleles will always be expressed. Each daughter must receive her father's X chromosome and each son must receive his father's Y chromosome. Hence fathers cannot transmit X-linked genes to their sons.

The genes on the sex chromosomes are distributed unequally to males and females within families. This inequality produces characteristic patterns of inheritance with marked discrepancies in the numbers of affected males and females.

The pedigree pattern depends upon which sex chromosome carries the mutant gene and whether the trait is recessive or dominant. Occasionally these pedigree patterns may be mimicked by autosomal traits which show sex limitation and the distinguishing features are summarised in Table 7.4. If the affected males of an autosomal dominant trait with sex limitation are infertile then the pedigree pattern in identical to an X-linked recessive where males do not reproduce. In this event the demonstration of lyonisation in carrier females is an important clue to the correct mode of inheritance.

Y-LINKED INHERITANCE (Holandric inheritance)

The inheritance of the cell surface marker detected by the monoclonal antibody 12E7 provides a human example of Y-linked inheritance. This marker is determined by the activity of a gene on the short arm of the X chromosome, MIC2, which may be closely related to the gene for the Xg red cell antigen. However, males with a deleted MIC2 gene may still express the 12E7 antigen due to the presence of an active homologous gene on the Y chromosome. Males carrying this gene transmit it to their sons but not to their daughters. So far no human examples of Y-linked diseases have been established.

X-LINKED RECESSIVE INHERITANCE

Severe sex-linked muscular dystrophy (Duchenne muscular dystrophy) is an example of an X-linked recessive trait. This condition produces a progressive proximal muscle weakness with massive elevations of all muscle enzymes including creatine phosphokinase (CPK). The onset is in early childhood and most children are chairbound by 10 years of age and die of intercurrent infection by 20 years. Fig. 7.1 shows a pedigree from an affected family. This pedigree illustrates the typical features of X-linked recessive inheritance. There is a marked discrepancy in the sex ratio with only boys affected. The affected boys have a similar disease course (no variation in expression). Heterozygous females are clinically unaffected (carriers) but transmit the condition to the next generation. This results in a 'knight's move' pedigree pattern of affected males. The condition is never transmitted by an unaffected male.

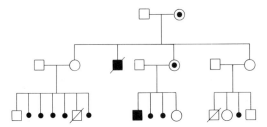

Fig. 7.1 Pedigree of a family with severe X-linked muscular dystrophy

Fig. 7.2 is a diagram of the expected proportions of affected to unaffected offspring when a carrier female marries a normal male. On average one half of her daughters will be carriers and one half of her sons will be affected. These proportions may be compared with the observed distribution in children of known carrier females using the chi-square test.

The carrier female is usually clinically normal. However, because of lyonisation a proportion of her muscle cells will have the active X which carries the mutant allele. These cells will release CPK and so

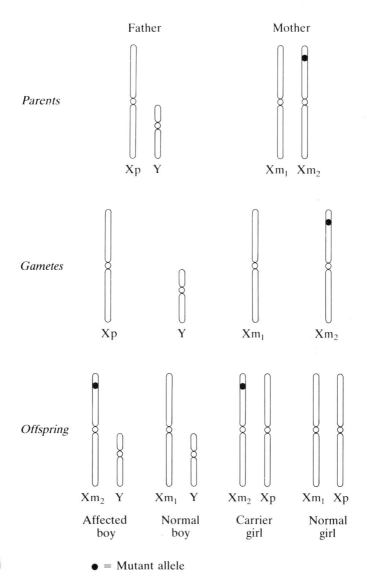

● = Mutant allele

Fig. 7.2 Diagram of the expected proportions of offspring for a female X-linked recessive heterozygote

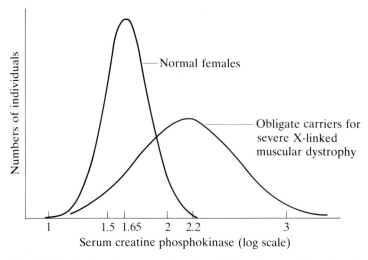

Fig. 7.3 Distribution of serum creatine phosphokinase in normal females and obligate carriers for severe X-linked muscular dystrophy

in about two-thirds of female carriers the CPK lies outside the normal range (Fig. 7.3). This is helpful in carrier detection provided precautions are taken to exclude other factors which can raise (exercise, intramuscular injections) or lower (pregnancy) this enzyme level. A woman with an affected child and an affected brother or a woman with more than one affected child is an obligate carrier as the alternative explanation of multiple new mutations is so unlikely. For each daughter of an obligate carrier there is a one in two risk that she too is a carrier. CPK testing may help to resolve this and carrier detection using linked restriction fragment length polymorphisms (RFLPs) may soon be applicable. Bayes' theorem is used to combine this information about the carrier risk (Appendix II).

Sometimes the child is the only affected individual in the family (Fig. 7.4). In this situation the mother is not an obligate carrier. In

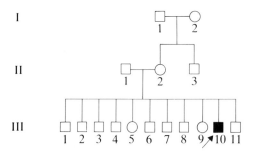

Fig. 7.4 Pedigree of a family with only one child affected by severe X-linked muscular dystrophy

perhaps one-third of cases the child has a new mutation whereas in the remainder the mother is a carrier (Chapter 10). CPK testing may help to resolve these two possibilities and the fact that this woman also has eight normal sons will diminish (but not obviate) her chance of being a carrier (Appendix II).

Occasionally a female is affected with this X-linked form of muscular dystrophy. This might arise in a number of ways:
1. Atypical lyonisation (manifesting heterozygote)
2. New mutation on the other X chromosome of a carrier female
3. A carrier with Turner's syndrome (45,X)
4. X-autosome translocation

By far the commonest of these possibilities is atypical lyonisation. This arises by the chance inactivation in muscle cells of most of her normal X chromosomes. Such a manifesting heterozygote is usually not affected to the same extent as a hemizygous male.

Theoretically a carrier female might also have a new mutation at the same locus on her other X chromosome and she would then be as severely affected as a male. A woman with 45,X or monosomy for the short arm of the X cannot inactivate her normal X chromosome which carries the mutant allele and so she would be as severly affected as a male. Finally, a woman with an X-autosome translocation may be affected. In X-autosome translocations the normal X is preferentially inactivated as otherwise partial monosomy for the involved autosome would occur. Eleven females are known with X-autosome translocations and severe X-linked muscular dystrophy. In each case the breakpoint was in the band Xp21 and this presumably resulted in damage to the gene for X-linked muscular dystrophy with consequent expression of this abnormal gene when the normal X was inactivated. This provided an important clue to the localisation of the gene for X-linked muscular dystrophy.

This severe form of X-linked muscular dystrophy is called Duchenne muscular dystrophy. There is also a milder X-linked form of muscular dystrophy called Becker muscular dystrophy. Recent linkage studies have provided evidence that these conditions might be due to different mutant alleles at the same locus (multiple allelism). Before counselling a family with muscular dystrophy it is important to establish the precise type as in addition to these X-linked forms autosomal dominant and recessive variants are known (genetic heterogeneity).

Other X-linked recessive conditions

So far 255 X-linked recessive traits are known in man. Some of the commoner and more clinically important of these are listed in Table 7.1. The frequencies vary in different ethnic groups, for example, colourblindness is rare in Eskimoes and in certain ethnic groups

Table 7.1 Human X-linked traits

Trait	UK frequency /10,000 males
Red-green colourblindness	800
Fragile X mental retardation	5
Non-specific X-linked mental retardation	5
Duchenne muscular dystrophy	3
Becker muscular dystrophy	0.5
Haemophilia A (factor VIII)	2
Haemophilia B (factor IX)	0.3
X-linked ichthyosis	2
X-linked agammaglobulinemia	0.1

glucose 6-phosphate dehydrogenase deficiency is as frequent as colourblindness in the UK (Chapter 16).

For some X-linked recessive disorders affected males may reproduce and in this event all daughters will be carriers (obligate carriers) and all sons will be normal (Fig. 7.5).

X-LINKED DOMINANT INHERITANCE

The Xg blood group serves to illustrate an X-linked dominant trait. This gene is located near the tip of the short arm of the X chromosome. Use of an antiserum distinguishes two types of individuals, Xg positive and Xg negative. These phenotypes are determined by a pair of alleles at the Xg locus termed Xga and Xg. Hence in the male two genotypes are possible whereas in the female three are found (Table 7.2). The female dominant homozygote cannot be distinguished from the heterozygote with the available methods of testing. Fig. 7.6 shows a pedigree with the Xg blood group marked under each person and those with Xg(a+) shaded.

Superficially this resembles the pedigree for an autosomal dominant trait but the critical difference lies in the offspring of the Xg(a+) man: all of his daughters are Xg(a+) but his sons are all Xg(a-). For a heterozygous mother married to an Xg(a-) father one half of her daughters and one half of her sons will be Xg(a+).

Overall an X-linked dominant disorder will be more frequent in females than males reflecting the relative distribution of sex chromosomes.

Table 7.2 Xg Blood group phenotypes and genotypes

Male		Female	
Genotypes	Phenotypes	Genotypes	Phenotypes
Xg^a Y	Xg(a+)	Xg^{a}Xg^a	Xg(a+)
		Xg^{a}Xg	
Xg Y	Xg(a−)	XgXg	Xg(a−)

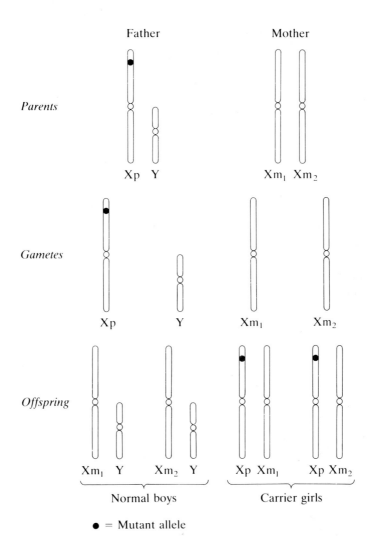

Fig. 7.5 Diagram of the expected proportions of offspring for an affected male with an X-linked recessive trait

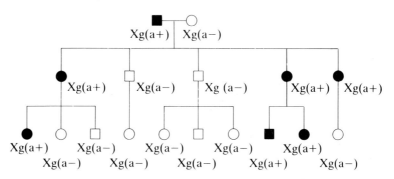

Fig. 7.6 Pedigree of a family to demonstrate inheritance of the Xg blood group. Phenotypes are indicated for each individual and those with Xg(a+) are shaded

Other X-linked Dominant Traits

There are 9 known human X-linked dominant traits (Table 7.3). With the exception of the Xg blood group, all are rare. For X-linked dominant diseases, males have a uniformly severe disease whereas the female heterozygotes tend to be more variably affected because

Table 7.3 Examples of Human X-linked dominant traits

Xg blood group
Pseudohypoparathyroidism
Vitamin-D resistant rickets
Incontinentia pigmenti*
Rett syndrome*

* lethal in hemizygous male

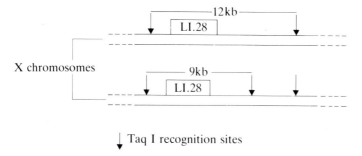

↓ Taq I recognition sites

Fig. 7.7 Diagram of two X chromosomes, one with and one without the extra recognition site for the enzyme Taq I. The probe L1.28 will hybridise to a 12 kb fragment from the upper chromosome and to a 9 kb fragment from the lower

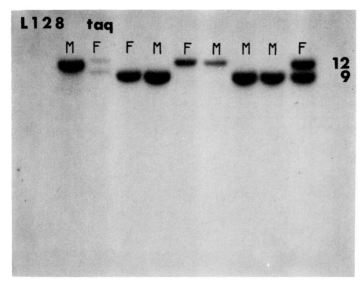

Fig. 7.8 Autoradiograph of separated 12 kb and 9 kb fragments after digestion with Taq I and hybridisation to L1.28. Sex of each individual is indicated.

of lyonisation. In two of these conditions (indicated by an asterisk) the affected males are so severely affected that spontaneous abortion is usual and there is then a marked excess of affected females.

X-LINKED CODOMINANT TRAITS

Restriction fragment length polymorphisms (RFLPs) provide several examples of X-linked codominant traits (ie. both alleles can be identified in the heterozygote).

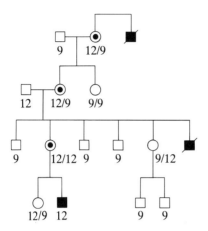

Fig. 7.9 Family showing codominant inheritance of the restriction fragment length polymorphism described in Figs. 7.7 and 7.8. Shaded individuals have severe X-linked muscular dystrophy

Table 7.4 Comparison of autosomal dominant with sex-limitation to X-linked recessive and dominant inheritance

	X-linked recessive	X-linked dominant	Autosomal dominant with sex-limitation
Pedigree pattern	Knight's move	Vertical	Vertical
Sex ratio	M > > F	2F : 1M	M > F
Male-to-Male transmission	Never	Never	50% of sons affected
Male-to-female transmission	All daughters carriers	All daughters affected	< 50% of daughters affected
Female-to-female transmission	50% of daughters carriers	50% of daughters affected	< 50% of daughters affected
Male severity	uniform	uniform	variable
Female severity	mild	variable	variable

One of these RFLPs is identified by a unique sequence X probe known as L1.28. This probe maps to the short arm of the X chromosome close to the centromere. With the restriction enzyme Taq I, 33% of males produce a single 9kb band and the remainder produce a 12 kb band as they lack a recognition site for this enzyme (Fig. 7.7). Females with two X chromosomes may be homozygous 12kb, heterozygous or homozygous 9kb (Fig. 7.8). The inheritance of this codominant marker is shown in a family in Fig. 7.9.

SUMMARY

Table 7.4 compares the main features of X-linked and autosomal sex-limited inheritance.

Chapter 8
Gene Mapping

Pedigree pattern analysis may confirm that a disorder is inherited as a single gene autosomal or sex-linked trait (Chapters 6 and 7). The next step is to attempt to identify the locus or position on the chromosome of the mutant gene. Such gene mapping is a prerequisite for improvements in prenatal diagnosis and carrier detection for the disorder. Furthermore, when the nature of the mutation is determined this should act as a rational basis for therapy. Gene mapping also provides an improved understanding of the detailed anatomy of the chromosomes; gives information on the parental origin of new mutations; and by comparison with other species is useful in the study of evolution.

There are four main methods for human gene mapping:
1. Family linkage studies
2. Gene dosage methods
3. In situ hybridization
4. Interspecific somatic cell hybridization

1. FAMILY LINKAGE STUDIES

The confirmation that a disorder is linked to a particular sex automatically assigns or locates the causative gene to the appropriate sex chromosome. Thus 12E7 may recognise a gene on the Y chromosome and Duchenne muscular dystrophy and the Xg blood group are determined by genes on the X chromosome. Chromosomal localisation is much more difficult for autosomal single gene disorders and in any event chromosomal localisation is only the first step in precise gene mapping.

Family linkage studies provided the earliest information about regional gene mapping in man and are currently increasing in importance as more chromosomal restriction fragment length polymorphisms are discovered.

Two genes are linked if their *loci* are close together on the same chromosome. In this situation the alleles at these two loci tend to pass together rather than independently into each gamete. Thus

Table 8.1 Marker traits for family linkage studies

Chromosomal heteromorphisms
Blood groups
Serum protein polymorphisms
Immunogenetic polymorphisms
Serum enzyme polymorphisms
Red cell enzyme polymorphisms
Restriction fragment length polymorphisms

disturbance of independent assortment (Mendel's second law) is an important clue that the two genes are linked. If the chromosomal location of one of the genes is known then by inference the other can be mapped to that area of the chromosome.

In a family linkage study two loci are considered, one for the disease or trait in question and another, the marker. A wide variety of marker traits may be used (Table 8.1). Each family member is examined to determine whether or not they are affected and to assess their status with respect to the marker trait. If the disease and the marker loci are on separate chromosomes then independent assortment will occur and the disease and marker should be found as often together as apart in the gametes and hence the offspring (Fig. 8.1). If, however, the disease and marker loci are close together on the same chromosome then independent assortment will not occur and the disease and marker will occur together in each child unless by chance they are separated by a crossover at meiosis (Fig. 8.2). On average there are about 52 crossovers in total at male meiosis with 1

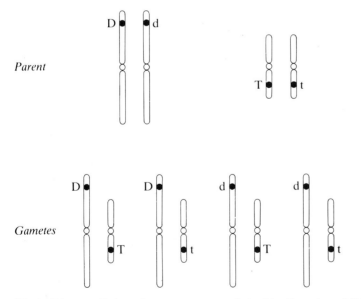

Fig. 8.1 Diagram of independent assortment at meiosis of the disease locus (alleles D and d) and the marker locus (alleles T and t)

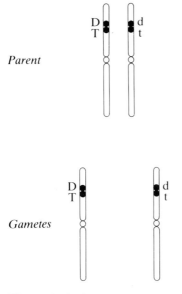

Fig. 8.2 Lack of independent assortment at meiosis for tightly linked disease locus (alleles D and d) and marker locus (alleles T and t)

to 6 per chromosome in proportion to the length of each. If the disease and marker loci are far apart on the same chromosome a crossover between the loci is very likely: the disease and marker traits will occur separately in each recombinant but together in the non-recombinants (Fig. 8.3). So for a distant marker trait the number of recombinants will equal the number of non-recombinants and the recombination fraction (number of recombinants divided by the total number of offspring) will be 0.5. This mimics independent assortment. As the distance between the marker and disease loci *decreases* the chance of crossing over diminishes, the number of recombinants becomes fewer, the recombination fraction falls, and the disturbance of independent assortment increases (Table 8.2). Thus the recombination fraction varies from 0 (tight linkage) to 0.5 (equivalent to independent assortment).

If a disease and marker loci are on different chromosomes then just by chance in a family each child who receives the chromosome with the marker may receive the chromosome with the disease trait. Obviously this becomes increasingly unlikely as more offspring are studied. A measure of the statistical significance of the departure from expected is given by the logarithm of the odds (Lod score) of observing such a disturbance of independent assortment by chance alone. A Lod score of 3 thus means that the observed disturbance would occur by chance only once in 10^3 or 1000 times. With this level of significance the alternative hypothesis, that the loci are in fact linked, is accepted.

Consider the nail-patella syndrome. This is inherited as an

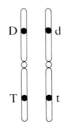

Parent

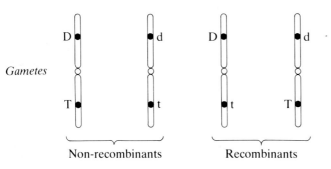

Gametes

Non-recombinants   Recombinants

Fig. 8.3 Distant disease and marker loci on the same chromsome mimic independent assortment at meiosis due to frequent crossovers

Table 8.2 Comparison of linked and unlinked loci

	Situation of the two loci			
	Same chromosome			Different chromosome
	very close	near	far apart	
Independent assortment	none	some	normal	normal
Number of recombinants	none	some	50%	50%
Recombination fraction	o	0–0.49	0.5	0.5

autosomal dominant trait and results in absent patellae, poor nails and kidney problems in adulthood. The pedigree of an affected family is shown in Fig. 8.4 together with the ABO blood group of each individual. Without exception each child who received the disease allele received the father's A allele and each who received the normal allele received his O allele. In this family there is a marked disturbance of independent assortment for the disease, nail-patella syndrome, and the marker, the ABO blood group. The likelihood of observing such a disturbance of independent assortment by chance alone is $(1/2)^7$ or 128 to 1 (Lod score 2.107).

In contrast, consider the same family using a different marker the

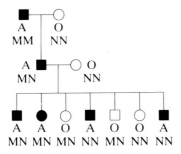

Fig. 8.4 Pedigree of a family with the nail-patella syndrome. ABO and MN blood groups are indicated for each individual

MN blood group. The MN blood group is determined as an autosomal codominant trait, so for the two alleles M and N there are three possible genotypes MM, MN and NN. Here the children who receive the father's mutant allele are equally likely to receive his M or his N allele (i.e. independent assortment). This excludes linkage between the MN blood group and the disease.

The suspected linkage with the ABO blood group was confirmed by further family linkage studies with an overall recombination fraction of 0.1 (10%). When the ABO blood group was assigned to 9qter by gene dosage studies with adenylate kinase 1 this automatically localised the gene for the nail-patella syndrome to the same area. The locus for the MN blood group is now known to be on chromosome 4.

Consider a further family with the nail-patella syndrome whose ABO blood groups are known (Fig. 8.5). Although these loci are linked this family provides no information to either support or

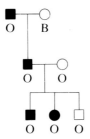

Fig. 8.5 Pedigree of a family with the nail-patella syndrome. ABO blood groups are indicated for each individual.

refute the fact as both parents have the same alleles at the marker locus. Such non-informative families pose an important practical problem in linkage studies whether for research or prenatal diagnosis.

In seeking to establish autosomal linkage the ideal mating is a double backcross which has produced multiple children. This is a

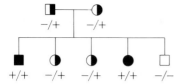

Fig. 8.6 Linkage of the locus for beta thalassaemia to a restriction endonuclease recognition site (+)

mating where one partner is heterozygous for both the disease and marker loci and the other is homozygous at both loci.

Linkage studies can also be used for autosomal recessive disorders. Fig. 8.6 shows the pedigree of a family with beta thalassaemia. This is inherited as an autosomal recessive trait and produces a severe anemia due to disturbed haemoglobin production. The heterozygous carriers are clinically normal but can be diagnosed by haematological testing. For each individual a recognition site for a restriction endonuclease on each homologue of chromosome 11 next to the beta globin gene is indicated as present + or absent -. For this site a person may be +/+, +/- or -/-. As in this family this site is tightly linked to beta thalassemia and in each heterozygous parent the chromosome 11 with the extra site is the homologue with the disease allele.

This situation where the marker (+) and the disease allele are on the same homologue is called linkage in coupling. If the marker (+) had been on the opposite homologue from the disease allele then the two are still linked (a property of their loci) but are now said to be linked in repulsion.

Assignment of a gene to a sex chromosome by the characteristic pedigree pattern is only the first stage in gene mapping of the X and Y. As no disease loci are yet assigned to the Y most interest has focussed on the X chromosome. Until the 1970's the only known X chromosome marker traits were colourblindness and the Xg blood group. Linkage was sought between these markers and each new X-linked disease. However, most families were not informative and only a few conditions were shown to be linked to these two markers (Table 8.3). In recent years restriction fragment length polymorphisms for the X chromosome have been identified. To date more than 20 of these are known for the X chromosome and these can be used as X marker traits in linkage analysis. In any family, although

Table 8.3 Linkage groups on the X chromosome

Linked to Xg — X-linked ichthyosis
　　　　　　 ocular albinism
Linked to colourblindness — Glucose 6-phosphate dehydrogenase
　　　　　　　　　　　　　　 Haemophilia A
　　　　　　　　　　　　　　 Fragile X mental retardation

some will be uninformative, others will provide data for or against linkage.

Due to the distribution of sex chromosomes in males and females, independent assortment of X chromosomes only occurs in the female. Hence the marker and disease traits are studied in the boys of obligate carrier females. The maximally informative family is where a mother heterozygous for both the marker and disease traits has numerous sons and is married to an unaffected father.

Using the restriction fragment length polymorphism identified by the probe L1.28 as the marker and the disease Duchenne muscular dystrophy there is a lack of independent assortment in the family shown in Fig. 7.9. The gene for Duchenne muscular dystrophy is assorting in this family with the 12kb fragment. Linkage between these loci was suspected and is now established with a recombination fraction of 0.15 (15%).

Using the same RFLP, linkage has also been established to the locus for Becker muscular dystrophy with a similar recombination fraction. The similarity of the distance as measured by recombination fraction between L1.28 and the Duchenne and Becker forms of muscular dystrophy raises the possibility that the two conditions might be due to different mutant alleles at the same locus.

RFLPs are not only important tools for gene mapping but are also now in clinical application for carrier detection and prenatal diagnosis of several conditions including beta thalassaemia, phenylketonuria, Huntington's chorea and haemophilia.

Recombination fraction and map distance

The relationship between recombination fraction and the actual physical distance between loci depends upon several factors. A recombination fraction of 0.1 (10%) indicates a map distance of 10 centimorgans (100 centimorgans = one Morgan) but with increasing distance apart the apparent recombination fraction falls due to the occurrence of double crossovers (Fig. 8.7). Furthermore crossovers for autosomes are more frequent in females than males and crossing over frequency also varies in different parts of the chromosome and seems to be greater at the ends that near the centromere.

Thus the physical distance represented by a given recombination fraction needs to incorporate all of these factors. The total length of the haploid genome is 3000 centimorgans and as the DNA therein has 3×10^9 base pairs on average one centimorgan equals one million base pairs.

In practice it is difficult to pick up linkages for loci more then 25 centimorgans apart and beyond 50 centimorgans frequent crossovers mimic independent assortment (recombination fraction 0.5).

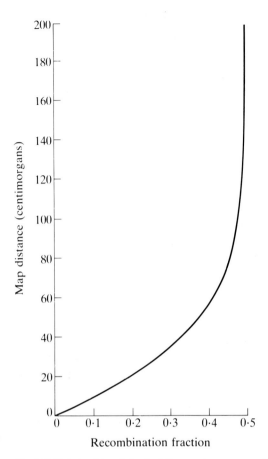

Fig. 8.7 Relationship of physical distance between two loci and the frequency of crossing-over (recombination fraction)

If three loci on the same chromosome are considered then the relative recombination fractions indicate the order of the genes.

2. GENE DOSAGE METHODS

Autosomal genes are arranged in pairs and both alleles are normally expressed. If two normal alleles are present for an enzyme then the activity of that enzyme will be 100%. Unbalanced autosomal chromosomal aberrations can provide information about genes carried by these chromosomes from dosage studies. If the area of the autosome which contains the enzyme locus is deleted then the residual enzyme activity will be 50% reflecting the activity of the remaining normal allele. Conversely if the person is trisomic for that area then the enzyme level will be 150%. By comparing enzyme levels in a variety of aberrations with different breakpoints the

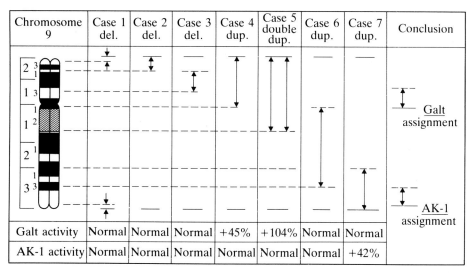

Chromosome 9	Case 1 del.	Case 2 del.	Case 3 del.	Case 4 dup.	Case 5 double dup.	Case 6 dup.	Case 7 dup.	Conclusion
Galt activity	Normal	Normal	Normal	+45%	+104%	Normal	Normal	
AK-1 activity	Normal	Normal	Normal	Normal	Normal	Normal	+42%	

Fig. 8.8 Application of gene dosage to localise galactose-1-phosphate uridyl transferase (GALT) and adenylate kinase 1 (AK-1) to particular areas of chromosome 9

location of the gene can be determined . This technique was first used to confirm the assignment of acid phosphatase to the short arm of chromosome 2, suggested by the observation that a child with a deletion of chromosome 2 had failed to inherit his mother's acid phosphatase allele. Fig. 8.8 shows how this method was used to locate two other red cell enzyme loci to particular segments of chromosome 9.

3. IN SITU HYBRIDISATION

The principle of in situ hybridization is to use DNA from the gene of interest to identify within the chromosome complement its complementary segment. The gene's DNA may be constructed by cDNA cloning from mRNA or by isolation from a whole genomic or chromosome-specific library (Chapter 2). This DNA is radiolabelled and the area of the chromosome to which it hybridises is revealed by autoradiography using standard chromosome preparations.

This technique was first used to map the immunoglobulin kappa light chain gene to the short arm of chromosome 2.

4. INTERSPECIFIC SOMATIC CELL HYBRIDISATION

A somatic cell hybrid is the fusion product of two somatic cells. For gene mapping one cell is of human origin and one of a different species for example a mouse. Initially such a hybrid will have a

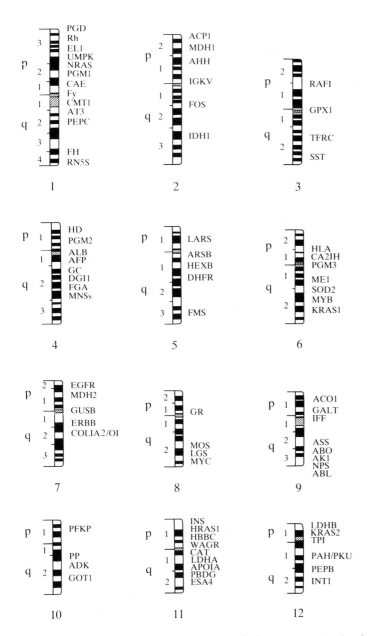

Fig. 8.9 Some more important assignments to the human gene map (see key)

ABL	9	Onc gene: Abelson strain of murine leukemia virus
ABO	9	ABO blood group
ACO1	9	Aconitase, soluble
ACP1	2	Acid phosphatase-1
ADA	20	Adenosine deaminase
ADK	10	Adenosine kinase
AFP	4	Alpha-fetoprotein
AHH	2	Aryl hydrocarbon hydroxylase
AK1	9	Adenylate kinase-1 (soluble)
ALB	4	Albumin
APO1A	11	Apolipoprotein A-I
APRT	16	Adenine phosphoribosyltransferase
ARSA	22	Arylsulfatase A
ARSB	5	Arylsulfatase B
ASS	9	Argininonsuccinate synthetase
AT3	1	Antithrombin III
B2M	15	Beta-2-microglobulin
BMD	X	Becker muscular dystrophy
C3	19	Complement component-3

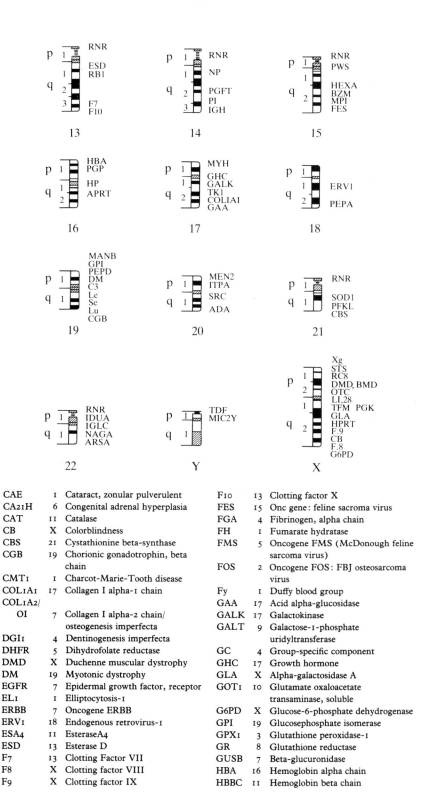

CAE	1	Cataract, zonular pulverulent
CA21H	6	Congenital adrenal hyperplasia
CAT	11	Catalase
CB	X	Colorblindness
CBS	21	Cystathionine beta-synthase
CGB	19	Chorionic gonadotrophin, beta chain
CMT1	1	Charcot-Marie-Tooth disease
COL1A1	17	Collagen I alpha-1 chain
COL1A2/ OI	7	Collagen I alpha-2 chain/ osteogenesis imperfecta
DGI1	4	Dentinogenesis imperfecta
DHFR	5	Dihydrofolate reductase
DMD	X	Duchenne muscular dystrophy
DM	19	Myotonic dystrophy
EGFR	7	Epidermal growth factor, receptor
ERBB	7	Oncogene ERBB
ERV1	18	Endogenous retrovirus-1
ESA4	11	EsteraseA4
ESD	13	Esterase D
F7	13	Clotting Factor VII
F8	X	Clotting factor VIII
F9	X	Clotting factor IX

F10	13	Clotting factor X
FES	15	Onc gene: feline sarcoma virus
FGA	4	Fibrinogen, alpha chain
FH	1	Fumarate hydratase
FMS	5	Oncogene FMS (McDonough feline sarcoma virus)
FOS	2	Oncogene FOS: FBJ osteosarcoma virus
Fy	1	Duffy blood group
GAA	17	Acid alpha-glucosidase
GALK	17	Galactokinase
GALT	9	Galactose-1-phosphate uridyltransferase
GC	4	Group-specific component
GHC	17	Growth hormone
GLA	X	Alpha-galactosidase A
GOT1	10	Glutamate oxaloacetate transaminase, soluble
G6PD	X	Glucose-6-phosphate dehydrogenase
GPI	19	Glucosephosphate isomerase
GPX1	3	Glutathione peroxidase-1
GR	8	Glutathione reductase
GUSB	7	Beta-glucuronidase
HBA	16	Hemoglobin alpha chain
HBBC	11	Hemoglobin beta chain

double chromosomal complement with both human and mouse chromosomes. The human chromosomes are selectively lost: rapidly at first and then more gradually. Mouse and human chromosomes can be distinguished on the basis of their different morphology and by staining with Giemsa at alkaline pH when the mouse chromosomes are bright purple whereas the remaining human chromosomes are pale blue.

HD	4	Huntington disease	NAGA	22	N-acetyl-alpha-D-galactosaminidase
HEXA	15	Hexosaminidase A	NP	14	Nucleoside phosphorylase
HEXB	5	Hexosaminidase B	NPS	9	Nail-patella syndrome
HLA	6	Human leukocyte antigens	NRAS	1	Oncogene NRAS
HP	16	Haptoglobin	OTC	X	Ornithine transcarbamylase
HPRT	X	Hypoxanthine-guanine phosphoribosyltransferase	PAH/ PKU	12	Phenylalanine hydroxylase/ phenylketonuria
HRAS1	11	Harvey rat sarcoma-1 protooncogene	PBDG	11	Porphobilinogen deaminase
IDH1	2	Isocitrate dehydrogenase soluble	PEPA	18	Peptidase A
IDUA	22	Alpha-l-iduronidase	PEPB	12	Peptidase B
IFF	9	Interferon, fibroblast	PEPC	1	Peptidase C
IGH	14	Immunoglobulin heavy chain gene cluster	PEPD	19	Peptidase D
			PFKL	21	Phosphofructokinase, liver
IGKV	2	Gene (cluster) for kappa light chain	PFKP	10	Phosphofructokinase, platelet
IGLC	22	Gene (cluster) for lambda light chain	PGD	1	6-Phosphogluconate dehydrogenase
			PGFT	14	Phosphoribosylglycinamide formyltransferase
INS	11	Insulin	PGK	X	Phosphoglycerate kinase
INT1	12	Oncogene INT: putative murine mammary cancer oncogene	PGM1	1	Phosphoglucomutase-1
			PGM2	4	Phosphoglucomutase-2
ITPA	20	Inosine triphosphatase	PGM3	6	Phosphoglucomutase-3
KRAS1	6	Kirsten rat sarcoma protooncogene-1	PGP	16	Phosphoglycolate phosphatase
			PI	14	Alpha-1-antitrypsin
KRAS2	12	Kristen rat sarcoma protooncogene-2	P	10	Inorganic pyrophosphatase
			PWS	15	Prader-Willi syndrome
L1.28	X	Unique sequence DNA probe	RAF1	3	Oncogene RAF1
LARS	5	Leucyl-tRNA synthetase	RB1	13	Retinoblastoma
LDHA	11	Lactate dehydrogenase A	RC8	X	Unique sequence DNA probe
LDHB	12	Lactate dehydrogenase B	Rh	1	Rhesus blood group
Le	19	Lewis blood group	RN5S	1	5S RNA gene(s)
LGS	8	Langer-Giedion syndrome	RNR	13–15, 21,22	Ribosomal RNA
Lu	19	Lutheran blood group			
MANB	19	Lysosomal alpha D-mannosidase	Se	19	Secretor
MDH1	2	Malate dehydrogenase, soluble	SOD1	21	Superoxide dismutase, soluble
MDH2	7	Malate dehydrogenase, mitochondrial	SOD2	6	Superoxide dismutase, mitochondrial
MEI	6	Malic enzyme	SRC	20	Oncogene SRC (Rous sarcoma)
MEN2	20	Multiple endocrine neoplasia, type II (Sipple syndrome)	SST	3	Somatostatin
			STS	X	Steroid sulfatase
MIC2	7	Surface marker recognised by monoclonal antibody 12E7	TDF	Y	Testis determining factor
			TFM	X	Testicular feminization syndrome
MNS	4	MN blood group	TFRC	3	Transferrin receptor
MOS	8	Onc gene: Moloney murine sarcoma virus	TK1	17	Thymidine kinase, soluble
			TPI	12	Triosephosphate isomerase
MPI	15	Mannosephosphate isomerase	UMPK	1	Uridine monophosphate kinase
MYB	6	Onc gene: avina myelobastosis virus	WAGR	11	Wilms tumor/aniridia/ gonadoblastoma/retardation
MYC	8	Onc gene: myelocytomatosis virus			
MYH	17	Myosin heavy chain	Xg	X	Xg blood group

If the mouse parent line used in the hybrid is homozygous for thymidine kinase deficiency the presence of this enzyme in a hybrid cell indicates that the human chromosome which produces this enzyme is present in that cell. By studying a selection of hybrids in this way it was possible to correlate the presence of thymidine kinase activity with the presence of human chromosome 17. Thus the locus for thymidine kinase was assigned to chromosome 17: this was the first human chromosomal assignment using somatic cell hybridisation.

NUMBERS OF ASSIGNMENTS

With the above approaches there has been a rapid growth in the number of human gene assignments (Table 8.4). Thus 247 autosomal genes and 118 sex-linked genes are now firmly assigned. This represents over 10% of the known human single gene traits. At least as many traits again have been provisionally assigned. As yet , however, relatively few of these mapped genes have been sequenced.

Table 8.4 Numbers of confirmed chromosomal assignments

Year	Autosomal assignments	X-linked assignments
1973	31	88
1974	48	91
1975	72	95
1977	83	102
1979	123	112
1981	180	116
1983	247	118

CURRENT GENE MAP

Fig. 8.9 shows the current gene map. It includes the major disease genes and genes of practical importance for family linkage studies. For genes not covered here see the reference section.

Certain generalisations are possible. Loci for enzymes in the same metabolic pathway are scattered, not clustered. Loci for lysosomal enzymes are scattered. Loci for mitochondrial and soluble forms of the same enzyme are scattered. Loci for subunits of complex proteins are not necessarily linked.

The map is similar in man and chimp thus reflecting the similarity in the banding pattern. The X chromosomal gene map appears similar in all mammals (Ohno's law) but other homologies diverge in proportion to the time of separation during evolution, mostly by the process of chromosomal rearrangement and unequal crossing-over.

Chapter 9
Multifactorial Inheritance

In contrast to single gene inheritance, either autosomal or sex-linked, the pedigree pattern does not allow a diagnosis of multifactorial inheritance. Single gene traits are discontinuous with mutant alleles producing distinct phenotypes. Multifactorial traits may be discontinuous or continuous but in each the trait is determined by the interaction of a number of genes at different loci each with a small but additive effect together with environmental factors. For discontinuous multifactorial traits, such as spina bifida, the risk within affected families is raised above the general population risk. The risk within an affected family is, however, low in comparison to single gene traits and rapidly falls towards the general population risk in more distant relatives. Thus in practice the proband with a discontinuous multifactorial trait is often the only affected person in that family.

For continuous multifactorial traits, such as height, there is only a range with continuous gradation between the two extremes. Most normal human characteristics are determined in this fashion.

In the analysis of a discontinuous trait it is first necessary to show that the incidence in members of affected families is increased above the general population incidence. If the incidence is not increased the condition is probably non- genetic. If increased, the pedigree pattern is examined for evidence of single gene inheritance. If multifactorial inheritance is then suspected and for the analysis of continuous traits studies of twin concordance and family correlation are necessary. These studies show that many congenital malformations and common diseases of adult life are inherited as multifactorial traits.

TWIN CONCORDANCE

Twins occur once in every 89 pregnancies. There are two types of twins:
1. Monozygotic (identical) - 33%
2. Dizygotic (non-identical) - 67%

Monozygotic twins arise from a single zygote which divides into two embryos during the first 14 days of gestation. Hence monozygotic twins have identical genotypes. Dizygotic twins result from two ova fertilized by two separate spermatozoa. Thus dizygotic twins have on average one half of their genes in commmon and are as similar genetically as siblings (brothers and sisters).

Diagnosis of zygosity

For research purposes the diagnosis of zygosity cannot be soley based upon a similar appearance. A record of the nature of the placental membranes and examination of polymorphic markers such as blood groups is necessary. All dizygotic twins have two amniotic sacs and two chorions. The chorions may secondarily fuse but the circulations of each part of the placenta always remain separate. The nature of the membranes for monozygotic twins depends upon the timing of zygotic division. In 75% of monozygotic twins there is a single chorion with a common placental circulation - this is diagnostic of monozygosity. In the remaining 25% two chorions are found and these cannot be distinguished from dizygotic twins by examination of placental membranes (Fig.9.1).

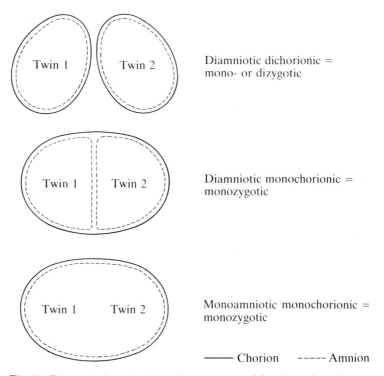

Fig. 9.1 Diagnosis of zygosity from the appearance of the placental membranes

If placental studies cannot confirm zygosity differences are sought between polymorphic markers in each twin.

Determination of concordance

Twins are concordant if they both show a discontinuous trait and discordant if only one shows the trait. There is a tendency to report concordant rather than discordant twin pairs and this bias needs to be considered in the interpretation of twin studies. For continuous traits the extent of the trait, for example height, is directly compared between the twins. As twins usually share a similar family environment it may be difficult to separate the relative extent of environmental and genetic contributions to a multifactorial trait. In this respect the concordance of monozygotic twins who have been reared apart from infancy is of major importance.

Results of twin studies

Monozygotic twins have identical genotypes whereas dizygotic twins are only as alike as siblings. If a condition has no genetic component, for example road traffic accidents, concordance rates are similar for both types of twins. For a single gene trait or a chromosomal disorder the monozygotic concordance rate will be 100% whereas the dizygotic rate will be less then this and equal to the rate in siblings. For multifactorial traits with both genetic and environmental contributions the rate in monozygotic twins, although less than 100% will exceed the rate in dizygotic twins (Table 9.1).

Tables 9.2 and 9.3 list the concordance rates for some continuous traits, congenital malformations and common adult disorders. In all of these multifactorial traits the concordance rate in monozygotic twins exceeds that in the dizygotic twins. The actual concordance rate in the monozygotic twins ranges from 6% to 95%. This range reflects the heritability of the condition. The higher the monozygotic concordance the more important the genetic contribution and so the higher the heritability.

Table 9.1

Disorder	Concordance	
	Monozygotic	Dizygotic
Single gene	100%	as sibs
Chromosomal	100%*	as sibs
Multifactorial	<100% but >sibs	as sibs
Environmental	as sibs	as sibs

* mosaicism excluded

Table 9.2 Twin concordance for continuous traits

Trait	Concordance	
	Monozygotic	Dizygotic
Height	95%	52%
IQ	90%	60%
Finger ridge count	95%	49%

Table 9.3 Twin concordance for some discontinuous traits

Trait	Concordance	
	Monozygotic	Dizygotic
Cleft lip	40%	5%
Spina bifida	6%	3%
Pyloric stenosis	15%	2%
Congenital dislocation of the hip	41%	3%
Talipes	23%	2%
Neural tube defect	6%	3%
Hypertension	30%	10%
Diabetes mellitus	55%	10%
Cancer	10%	10%
Epilepsy	37%	10%
Schizophrenia	60%	10%
Manic-depression	70%	15%
Mental retardation IQ < 50	60%	3%
Leprosy	60%	20%
Tuberculosis	51%	22%
Atopic disease	50%	4%
Hyperthyroidism	47%	3%
Psoriasis	61%	13%
Gallstones	27%	6%
Sarcoidosis	50%	8%

FAMILY CORRELATION

Relatives share a proportion of their genes (Table 9.4). Thus if a trait is determined by multifactorial inheritance, relatives should show the trait in proportion to their genetic similarity. This is really only an extension of the twin study technique. The similarity of different relatives in this respect is known as their correlation. Their correlation is measured on a scale of 0 to 1 where 1 is identical and 0 is as dissimilar as possible.

If parents are not blood relatives then they would be expected to be as alike genetically as random members of the population. Thus their correlation for genetically determined traits should be only equal to the general population average. In practice many slightly

Table 9.4

Degree of relationship	Examples	Proportion of genes in common
1st	parent to child, sib to sib	1/2
2nd	grandparent to grandchild, nephew or niece to aunt or uncle	1/4
3rd	first cousins	1/8

Table 9.5 Family correlations for some continuous traits

	Correlation of first degree relatives	
	Observed	Expected
Height	0.53	0.5
IQ	0.53	0.5
Finger ridge count	0.49	0.5

exceed this as a result of selective mating for such characteristics as height and intelligence (assortative mating - Chapter 10). The more closely related the relatives the higher the correlation should be for a genetically determined trait.

Table 9.5 shows the familial correlations for several continuous multifactorial traits. Height, intelligence, and total fingertip ridge count provide close family correlations to those predicted from the proportion of genes in common. Table 9.6 shows the frequency of some discontinuous traits in relatives of an affected person. The frequency falls off in proportion to the proportion of genes in common but is increased in all relatives above the general population frequency.

Thus twin concordance and family correlation studies support

Table 9.6 Frequency of discontinuous traits for differing degrees of relationship

Trait	Frequency			Population frequency
	1st degree relatives	2nd degree relatives	3rd degree relatives	
Cleft lip	4%	0.6%	0.3%	0.1%
Spina bifida	4%	2%	1%	0.8%
Pyloric stenosis	2%	1%	0.4%	0.3%
Epilepsy	5%	2.5%	1.5%	1%
Schizophrenia	10%	4%	2%	1%
Manic-depression	15%	5%	3.5%	1%

multifactorial inheritance for a trait, whether discontinuous or continuous.

In some disorders the nature of the interacting genetic and environmental components is now understood (Table 16.1). In each of these only a single locus interacts with the environmental agent(s) and so far little is known about the genetic or environmental components in more complex multifactorial disorders.

CONTINUOUS MULTIFACTORIAL TRAITS

Most normal human characteristics are determined as continuous multifactorial traits (Table 9.7). These traits by definition have a continuously graded distribution. Thus for intelligence there is a range from the severely subnormal to the most gifted with the average IQ of 100 (Fig. 9.2). As can be seen the distribution of intelligence in a population is Gaussian with the majority of

Table 9.7 Examples of human continuous multifactorial traits

Height
Weight
Intelligence
Total ridge count
Red cell size
Blood pressure
Skin colour

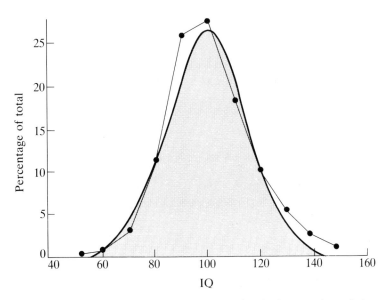

Fig. 9.2 Gaussian distribution of intelligence quotient in the general population

1 pair of alleles controlling a character

1 2 1

2 independent allelic pairs controlling
the same character

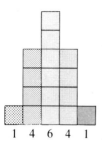

1 4 6 4 1

3 independent allelic pairs controlling
the same character

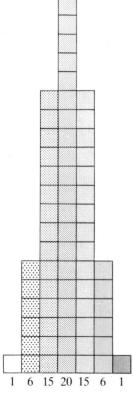

1 6 15 20 15 6 1

Fig. 9.3 - 9.5 Expected distribution of colour in offspring if the trait was due to a single locus with two alleles (Fig. 9.3), two loci each with two alleles (Fig. 9.4) and three loci each with two alleles (Fig. 9.5). Note the approach towards a Gaussian distribution.

individuals centred around the mean. Such a distribution is characteristic of continuous multifactorial traits.

The interaction of a number of loci to produce such a gradual range is readily appreciated from the range of shade produced by the interaction of pairs of alleles at one, two and three hypothetical loci for colour (Figs. 9.3-9.5).

Parents of above average intelligence tend to have children who are brighter than normal but who are not quite as bright as themselves. The mean IQ of the offspring will approximate to the mid-point between the mean parental IQ and the general population mean of 100. This is called regression towards the mean and also operates at the lower end of the normal range of intelligence. Regression towards the mean is more marked in families where the degree of parental deviation is extreme.

DISCONTINUOUS MULTIFACTORIAL TRAITS

More than 20 discontinuous multifactorial traits have been described in man. Table 9.8 lists some of the commoner of these which are of medical importance. Broadly these traits can be divided into congenital malformations and common conditions of adult life.

Cleft lip and palate is a congenital malformation inherited as a multifactorial trait (Fig. 9.6). In the mildest form the lip alone is unilateraly cleft whereas in the most severe the lip is bilaterally cleft and the palatal cleft is complete. The parents of this child were unaffected and there was no family history of cleft lip and palate. By virtue of having produced an affected child this indicates that each parent must have some underactive genes for lip and palate formation. However, as they have fully formed lips and palates they must also have on balance more normally active genes. For these discontinuous traits it is the critical balance between the number of underactive and the number of normally active genes which is important. Only when the balance exceeds a certain threshold will

Table 9.8 Discontinuous human multifactorial traits

Congenital malformations
   Cleft lip and palate
   Congenital heart disease
   Neural tube defect
   Pyloric stenosis

Common adult diseases
   Rheumatoid arthritis
   Epilepsy
   Peptic ulcer
   Schizophrenia
   Manic-depression

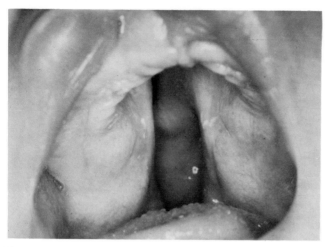

Fig. 9.6 Cleft lip and palate

the malformation occur and the further the threshold is exceeded the greater the extent of the malformation. Thus the liability (including both genetic and environmental factors) can be represented as a Gaussian curve (Fig. 9.7). The threshold is indicated and the proportion of the population to the right of this threshold (0.1%) equals the general population incidence of the condition. For parents (first degree relatives) of an affected child their liability curve is shifted to the right and so we would expect to find an increased frequency (4%) of this malformation amongst parents and other first degree relatives (Fig. 9.8). With each further degree of relationship the liability curve moves back a step towards the general population position with a corresponding reduction in the incidence (Table 9.6).

The more severe the malformation in the affected child the more the parents liability curve is shifted to the right and the higher the

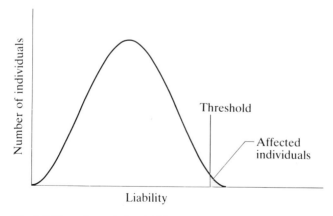

Fig. 9.7 General population liability curve for cleft lip and palate

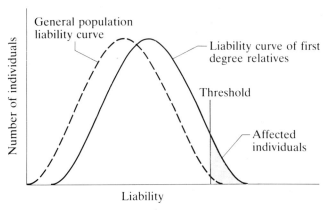

Fig. 9.8 Displaced liability curve in first degree relatives of a proband with cleft lip and palate

incidence in relatives. Thus 5% of first degree relatives are affected if the clefting is bilateral and complete whereas only 2% are affected if unilateral and incomplete.

Some multifactorial traits show a marked sex predilection (Table 9.9). Thus whilst pyloric stenosis affects 5/1000 males, only 1/1000 females are affected. The incidence is increased in the relatives of affected males but is even more increased in the relatives of affected females (Table 9.10). This indicates that the female threshold is higher than the male threshold for this malformation and so the parents of an affected female need to have a higher proportion of underactive genes and hence a more displaced liability curve.

Many common adult disorders are inherited as multifactorial traits. Whilst some such as peptic ulcer and rheumatoid arthritis are

Table 9.9 Multifactorial conditions with unequal sex ratio

Condition	Sex ratio (males to females)
Pyloric stenosis	5 to 1
Hirschsprungs disease	3 to 1
Congenital dislocation of the hip	1 to 6
Talipes	2 to 1
Rheumatoid arthritis	1 to 3
Peptic ulcer	2 to 1

Table 9.10 Pyloric stenosis frequency in relatives

	Male affected	Female affected
Incidence in sibs	2%	10%
Incidence in offspring	4%	20%

clearly discontinuous traits the distinction is less obvious for others such as diabetes mellitus and hypertension. There is obviously a range for both glucose tolerance and blood pressure and often the choice of upper limit of normal is a quite arbitrary point in an otherwise Gaussian distribution.

COMPARISON OF SINGLE GENE AND MULTI-FACTORIAL INHERITANCE

Although better understood, single gene inheritance is less important than multifactorial inheritance in the inheritance of both common diseases and normal characteristics. Table 9.11 summarises the salient differences.

Table 9.11 Comparison of single gene and multifactorial inheritance

	single gene	multifactorial
Pedigree	often diagnostic	non-diagnostic
Monozygotic twin concordance	100%	<100% but >sibs
Risk to relatives	often high	usually low
Frequency	rare, overall 1%	common, overall 15%

Chapter 10
Population Genetics

Population genetics is the study of the distribution of genes in populations and of how these gene frequencies are maintained or changed. The time scale for population geneticists is measured by the generation time of about 30 years (the mean age at the birth of offspring). This subject is relevant to all medical practitioners because of its implications for human evolution.

MAINTENANCE OF GENE FREQUENCIES

In a population the relative frequencies of different alleles tend to be maintained constant from one generation to the next. This can be demonstrated mathematically and helps to explain why dominant traits do not automatically increase at the expense of recessives.

Consider one locus with two alleles A and a. If the frequency of the allele A is p and the frequency of the allele a is q then since each invididual must have one or other allele the sum of these allele frequencies must be one or 100%.

Therefore $p + q = 1$

Fig. 10.1 shows the frequencies of each genotype at this locus.

In the production of the next generation each of the three types of paternal genotype may mate with each of the three types of maternal genotype (Fig. 10.2). Fig. 10.3 indicates the genotypes of the offspring for each mating type and as can be seen the relative frequencies of each is unchanged and the population is said to be in

Paternal gametes

		A (p)	a (q)
	A (p)	AA (p^2)	Aa (pq)
Maternal gametes	a (q)	Aa (pq)	aa (q^2)

Fig. 10.1

		AA (p^2)	Aa $(2pq)$	aa (q^2)
	AA (p^2)	AA x AA (p^4)	AA x Aa $(2p^3q)$	AA x aa (p^2q^2)
Maternal genotypes	Aa $(2pq)$	Aa x AA $(2p^3q)$	Aa x Aa $(4p^2q^2)$	Aa x aa $(2pq^3)$
	aa (q^2)	aa x AA (p^2q^2)	aa x Aa $(2pq^3)$	aa x aa (q^4)

Fig. 10.2

Mating type	Frequency (from Fig. 10.2)	Offspring AA	Aa	aa
AA x AA	p^4	p^4		
AA x Aa	$4p^3q$	$2p^3q$	$2p^3q$	
AA x aa	$2p^2q^2$		$2p^2q^2$	
Aa x Aa	$4p^2q^2$	p^2q^2	$2p^2q^2$	p^2q^2
Aa x aa	$4pq^3$		$2pq^3$	$2pq^3$
aa x aa	q^4			q^4

$$
\begin{aligned}
\text{AA offspring} &= p^4 + 2p^3q + p^2q^2 \\
&= p^2(p^2 + 2pq + q^2) = p^2(p+q)^2 = p^2(1)^2 = p^2 \\
\text{Aa offspring} &= 2p^3q + 4p^2q^2 + 2pq^3 \\
&= 2pq(p^2 + 2pq + q^2) = 2pq \\
\text{aa offspring} &= p^2q^2 + 2pq^3 + q^4 \\
&= q^2(p^2 + 2pq + q^2) = q^2
\end{aligned}
$$

Fig. 10.3

genetic equilibrium. Although the actual numbers of individuals with each genotype may have increased the relative proportions of each genotype (and allele) have remained constant (AA at p^2, Aa at $2pq$ and aa at q^2). This principle is called the Hardy-Weinberg law.

Applications of the Hardy-Weinberg Law

The most important application is the calculation of carrier frequencies for autosomal and X-linked recessive traits.

For any autosomal recessive trait, if q is the frequency of the mutant allele and p the frequency of the normal allele, then the frequency of the recessive homozygote is equal to the square of the mutant allele frequency (q^2).

Thus for cystic fibrosis:

Recessive homozygote frequency, $q^2 = 1/2000$

$q = \sqrt{1/2000} = 1/45$ or 0.0224

$p = 1 - q = 44/45$ or 0.9776

and so the heterozygote (carrier) frequency, $2pq = 1/23$ or 0.0438.

Table 10.1 Autosomal recessive disease and carrier frequencies

Disease	Birth frequency	Carrier frequency
Cystic fibrosis	1/2000	1/23
Congential deafness	1/5000	1/35
Phenylketonuria	1/10,000	1/50
Albinism (tyrosinase -ve)	1/40,000	1/100
Alkaptonuria	1/100,000	1/160

Table 10.2 Carrier female and affected male frequencies for X- linked recessive traits (UK data)

Trait	Affected male frequency	Carrier female frequency
Red-green colourblindness	1/12	1/7
Fragile X mental retardation	1/2000	1/1000
Non-specific X-linked mental retardation	1/2000	1/1000
Duchenne muscular dystrophy	1/3600	1/1800
Becker muscular dystrophy	1/20,000	1/10,000
Haemophilia A	1/5000	1/2500
Haemophilia B	1/30,000	1/15,000
X-linked ichthyosis	1/5000	1/2500

Table 10.1 indicates the homozygous affected and carrier frequencies for several autosomal recessive diseases. Thus for autosomal recessive traits most of the mutant alleles in the population are found in asymptomatic heterozygotes.

For X-linked recessive traits as the male is hemizygous (ie. a single X chromosome) the affected male frequency equals the frequency of the mutant allele (q).

So for X-linked recessive red-green colourblindness:

The affected male frequency , q = 1/12 or 0.08

Therefore p = 11/12 or 0.92

Coloublind females are homozygous for the recessive allele and so their frequency will be q^2 which is 0.0064 (0.64% or 1 in 156). The carrier frequency in females will be 2pq which is 0.14 (14% or 1 in 7). The ratio of affected males to affected females will be:

$q/q^2 = 1/q = 12$ affected males to each affected female

Table 10.2 indicates the frequencies of affected males and carrier females for several X-linked recessive traits of medical importance. Thus for X-linked recessive traits twice as many mutant alleles occur in carrier females as in affected males.

Disturbance of the gene frequencies in a population

Several factors can interfere with the gene frequencies in a population and lead to either an increase or a decrease in allele frequencies from one generation to the next. These factors are:
1. Non-random mating
2. Altered mutation rate
3. Selection
4. Small populations
5. Migration

1. Non-random mating

Random mating is the selection of a mate irrespective of the spouse's genotype. In practice mating is probably never entirely random as inherited factors such as height, weight, physique and intelligence tend to play a role (assortative mating).

Consanguinity or mating between genetic relatives is an example of non-random mating. Consanguineous individuals have at least one not-too-remote common ancestor. The offspring of these matings are at increased risk of homozygosity for any recessive alleles carried by the common ancestor(s). Generally such alleles are deleterious in the homozygous state and consanguinity will thus alter gene frequencies by the loss of two mutant alleles in each recessive homozygote who fails to reproduce. In general, for autosomal recessive traits the proportion of consanguineous parents increases in proportion to the rarity of the trait.

2. Altered mutation rate

A mutation is a change in the genetic material. The mutation rate (μ) is the frequency of such change and may be expressed as the number of mutations at that locus per million gametes produced. Most information about human mutation rates relates to autosomal dominant traits as it is more difficult to estimate the rate for recessive traits. For rare dominant traits the mutation rate may be calculated directly from the formula:

$$\mu = n/2N$$

where n is the number of affected patients with normal parents and N is the total number of births for the area and time period studied.

If a condition does not preclude reproduction then some new cases will have inherited the trait from an affected parent. The proportion of new mutations varies with the degree to which the condition hinders reproduction. If there is no impairment of reproductive fitness then the biological fitness (f) is normal or 1. If no reproduction is possible then the biological fitness is zero.

For an autosomal dominant trait at equilibrium the birth frequency is given by:

Equation I   Birth frequency $= 2\mu/(\text{1-f})$

where f is the biological fitness.

The birth frequencies, mutation rates and biological fitness for several autosomal dominant traits are given in Table 10.3. The birth

Table 10.3   Autosomal dominant trait birth frequencies in relation to mutation rate and biological fitness

Trait	Birth frequency	Mutations per million gametes	Biological fitness	New mutations
Neurofibromatosis	1/3000	100	0.4	25%
Marfan syndrome	1/66,000	5	0.4	15%
Achondroplasia	1/26,000	14	0.2	80%
Huntington's chorea	1/18,000	5	0.8	1%
Myositis ossificans	1/500,000	1	0.01	99%

frequency falls as the biological fitness decreases and is increased in proportion to the mutation rate. Thus the observed gene frequency represents a balance between loss from failure to reproduce (negative selection) and gain by new mutation.

For an autosomal recessive trait the birth frequency in a population at equilibrium is:

Equation II   Birth frequency $= \mu/(\text{1-f})$

Thus if the affected homozygote never reproduces (f=0) the birth frequency equals the mutation rate.

For an X-linked recessive trait the birth frequency in a population at equilibrium is:

Equation III   Birth frequency $= 3\mu/(\text{1-f})$

Thus for genetic lethals with a biological fitness of zero such as Duchenne muscular dystrophy the birth frequency equals 3μ. The female carrier frequency for an X-linked genetic lethal has been estimated to be 4μ. Thus the birth frequency represents the sum of 2μ from the carrier mother (half of 4μ as only one half of her sons are affected) plus μ (the chance of a new mutation for a son's X chromosome). Thus it is predicted that one third ($\mu/3\mu$) of isolated cases of Duchenne muscular dystrophy are new mutations and the remaining two thirds ($2\mu/3\mu$) are inherited from a carrier mother. Table 10.4 indicates some estimates for the mutation rate of X-linked recessive traits.

The risk of a new mutation for several autosomal dominant traits and for some X-linked recessive traits has been shown to be increased with increasing paternal age (Fig. 10.4). This may also be true for autosomal recessive traits but is more difficult to demonstrate.

Table 10.4 Estimates for the mutation rate for some X-linked recessive traits

Trait	Mutation rate (mutations per million gametes)
Haemophilia A	20–40
Haemophilia B	5–10
Duchenne muscular dystrophy	40–100

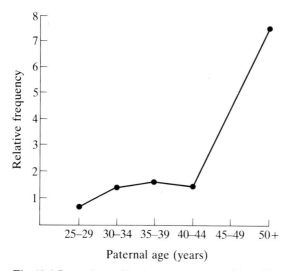

Fig. 10.4 Paternal age effect in progressive myositis ossificans

3. Selection

Selection is a powerful means for altering gene frequencies. It may operate to reduce (negative selection) or increase (positive selection) a particular phenotype and hence its genotype. Genetic selection acts on the individual phenotypes and either favours or hinders reproduction and thus the propagation of that individual's genotype.

Hence selection is acting by modifying an individual's biological fitness (f).

For an autosomal dominant trait (equation I in preceding section) any increase in the biological fitness, perhaps by improved treatment, will rapidly alter the gene frequency over the next few generations to a new equilibrium. Thus if the fitness was restored from 0% to 100% the frequency of the disorder would double in a generation and keep increasing by the same amount in later generations. For dominant conditions with relatively high initial fitness the increase in frequency would be much slower. With complete negative selection (f=0) the gene frequency would settle to 2μ.

For an autosomal recessive trait (equation II) selection against the recessive homozygote for instance by enforced sterilization is much less effective as most persons in the population with the gene are heterozygous carriers. Restoring the homozygote biological fitness from 0% to 100% would only double the gene frequency after 50 generations if the starting disease frequency was 1/15,000. If less frequent then the increase would be even slower. A reduction in homozygote biological fitness would result in a similar slow change in gene frequencies.

For an X-linked recessive trait (equation III) the situation is intermediate between autosomal dominant and recessive and an increase of biological fitness to 100% would double the birth frequency in about four generations.

Selection may also act on the recessive heterozygote as exemplified by sickle cell anaemia. Sickle cell anaemia is inherited as an autosomal recessive trait. The affected homozygotes have severe anaemia and despite therapy often die before adulthood. With such a low biological fitness the homozygote birth frequency might be expected to approach 1 in 100,000. In certain areas such as equatorial Africa, however, the birth frequency may be more than 1 in 40. Clearly the homozygote who dies without reproducing cannot have a selective advantage and so the advantage must lie with the heterozygote. The areas where sickle cell anaemia is most prevalent correspond geographically with the distribution of *Plasmodium falciparum* malaria (Fig. 10.5). This clue led to the discovery that in the heterozygote the red cells parasitised by Plasmodium falci-

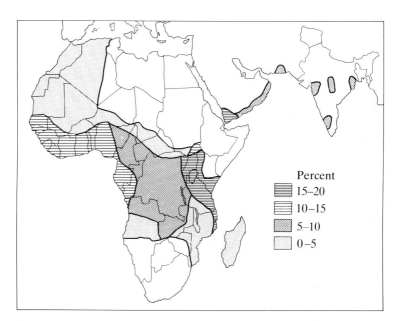

Fig. 10.5 Frequency distribution of the sickle cell gene

parum undergo sickling and are destroyed. The sickle carrier thus more readily overcomes malarial infection than the normal homozygote and is so at a reproductive advantage. Increased biological fitness in comparison with normal homozygotes leads to a relatively rapid change in gene frequencies.

This selective advantage no longer operates in regions where malaria has been eradicated. Thus the sickle cell gene frequency has fallen in the descendents of African negroes who were transported to America as slaves some 10 generations ago (Table 12.3). Heterozygotes for beta thalassaemia and glucose 6-phosphate dehydrogenase (G6PD) deficiency similarly have a selective advantage over homozygous normals by virtue of malarial resistance. Parasitisation of G6PD deficient cells does occur but at a much lower frequency than in the red cells with normal enzyme activity.

4. Small populations

For religious, geographical or other reasons a small group of individuals may become genetically isolated from the rest of the population (genetic isolate). The founder members of the group will be carriers of mutant alleles for some recessive traits and so within this population these genes are automatically at a higher frequency than within the general population (Table 10.5).

With only a small number of individuals in a breeding population the actual frequencies of alleles varies widely from one generation to the next (random genetic drift). By chance one allele may fail to be passed on to the next generation and so disappear (extinction) leaving only the alternative allele at that locus (fixation).

Table 10.5 Genetic isolates with high frequencies of certain autosomal recessive diseases

Isolate	Diseases
Amish	chondroectodermal dysplasia
Afrikaaners	variegate porphyria
Cuna Indians of Panama	albinism
Finns	congenital nephrotic syndrome
Eskimoes	congenital adrenal hyperplasia

5. Migration

Migrant individuals will modify the gene pool of their descendents. The best example of this is the gradual fall in frequency of the ABO allele B in moving westwards across Asia and Europe. Thus the frequency of B in Eastern Asia is 0.3 but in Western Europe is only

0.06. This would suggest that the B allele originally arose in Eastern Asia.

GENETIC POLYMORPHISMS

A genetic polymorphism is the occurrence together in the same population of two or more discontinuous traits at a frequency where the rarest could not be maintained by recurrent mutation alone. In general if 1 in 50 or more of the general population has the rare allele then the condition is polymorphic. Table 10.6 lists some examples of

Table 10.6 Examples of human polymorphic traits

Chromosomal
   (length of Yq, size of centromeric heterochromatin)
Blood groups
   (ABO, MN, Rh)
Cell surface antigens
   (HLA)
Red cell enzymes
   (adenylate kinase, phosphoglucomutase, acid phosphatase 1)
Serum proteins
   (Haptoglobins)
DNA endonuclease recognition sites

---

human polymorphisms. At least 30% of enzymes and proteins are polymorphic and this may well be an underestimate since only 75% of point mutations result in a changed aminoacid sequence and only about one third of proteins with altered aminoacid sequence have altered electrophoretic mobility. The defective enzymes and proteins associated with human disease are simply extreme examples of this type of variation.

A polymorphism may be produced by selection. Such selection is (or was) operating to increase the rare allele frequency by favouring the homozygote or the heterozygote. It has proved difficult to identify these selective pressures. They are probably small and may now be obsolete, for example certain infectious diseases such as smallpox. If obsolete then the rare allele frequency should gradually fall to a new equilibrium as has been seen for sickle cell anaemia in America (a changing or transient polymorphism). It has been suggested that some polymorphic mutations are neutral (ie. no advantage or disadvantage to the carrier). This is difficult to prove or disprove and in any case applies only to the present environment.

EVOLUTION

Prior to Darwin, the different species were held to be fixed since their outright creation. Darwin challenged this view by showing that traits favoured by the environment tended to increase in frequency as the favoured animals were more successful at reproduction. Conversely, deleterious traits hindered reproduction and thus fell in frequency. This natural selection operates on the phenotype which in turn is determined by the genotype. Thus evolution is simply a change in gene frequencies as a result of selection.

Life has existed on Earth for 3 billion years with mammals present for 70 million years and man only for 200-300,000 years (7-10,000 generations). Each species is a set of individuals who can interbreed and have fertile progeny. In general amongst mammals it may take in the order of one million years for two isolated populations to evolve from genetic identity to intersterility. As might be expected the genetic differences between related species (chromosome number, chromosome rearrangement, similarity of gene and hence protein structure) vary in proportion to the timing of evolutionary separation. For example, whilst all mammals have alpha globin chains the structure shows species variation (Table 10.7). Some portions of each protein show less variation than others.

Table 10.7 Number of aminoacid differences between four mammals for the alpha globin chain

	Man	Gorilla	Pig	Rabbit
Man	0	1	19	26
Gorilla		0	20	27
Pig			0	27
Rabbit				0

These conserved areas are usually vital for function and so preserved by natural selection.

Gene duplication appears to be a basic mechanism by which the genome can increase in size and permit more rapid and extensive protein evolution than can be achieved by mutation alone. There are already several human examples of duplicated genes which have diversified (eg. globin genes and collagen genes) and doubtless such gene families will be the usual pattern in as yet unmapped areas of the human genome.

Chapter 11
Immunogenetics

GENETICS OF THE NORMAL IMMUNE SYSTEM

The prime function of the immune system is to recognise and attack foreign (non-host) antigens. Most proteins, some polysaccharides and some nucleic acids are antigenic. The immune response has two components cellular and humoral (Fig. 11.1). In the humoral

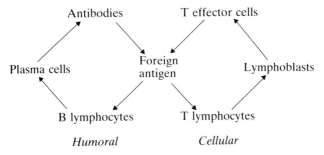

Fig. 11.1 Components of the immune response

response specific antibodies are produced by stimulated B lymphocytes or plasma cells. Antibodies are immunoglobulins and are important in the response to bacterial infections. The cellular immune response is effected by T lymphocytes which, when specifically stimulated, become T effector cells. This cellular response is important in the reaction to malignant cells, transplanted tissues, intracellular viruses and fungal infections.

Each immunoglobulin molecule is a protein composed of two identical light and two identical heavy chains held together by disulphide bonds (Fig. 11.2). Two types of light chain, kappa (κ) and lambda (λ) occur. The light chains are alike in all classes of immunoglobulin but each class has its own characteristic heavy chain (Table 11.1). Each immunoglobulin chain has three regions, a variable (V) region at the N terminal end which is part of the antibody combining site, a junctional (J) region and a constant (C) region.

The kappa light chain gene is on the short arm of chromosome 2

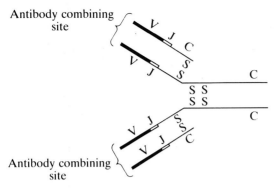

Antibody combining site

Antibody combining site

Fig. 11.2 Model of the immunoglobulin molecule

Table 11.1 Classes of immunoglobulin

Class	Molecular weight	Heavy chain	Light chain	Comment
IgG	150,000	γ	κ or λ	Abundant; only class to cross placenta; appears late in immune response
IgM	900,000	μ	κ or λ	Precedes IgG in immune response
IgA	160,000	α	κ or λ	Surface immunity
IgD	185,000	δ	κ or λ	Function unknown
IgE	200,000	ε	κ or λ	Allergic response

whereas the lambda light chain gene is on chromosome 22. The locus for the heavy chain gene is on chromosome 14. Each of these genes is actually a gene cluster. The heavy chain gene contains 50 to 150 variable portions with at least four junctional portions and a constant region for each class of immunoglobulin (Fig. 11.3). In each plasma cell, by a process of somatic recombination, only one V, one J and one C region are transcribed into a single contiguous molecule of messenger RNA (Fig. 11.4). This pattern will remain constant for that plasma cell and its descendents. A plasma cell which responds to a different antigen will express a different VJC combination and thus each plasma cell selects one allelic combination and excludes all others. This accounts for our ability to produce an estimated 10^7 antibodies of different specificities. This ability of a single gene cluster to produce a host of different polypeptides is an important exception to the one gene-one polypeptide rule. The kappa and lambda light chain genes are also clusters which produce diverse products by an identical mechanism (Fig. 11.5).

Complement is a series of plasma proteins which are required for the destruction of antigens. Most components of the complement series have loci close to the HLA complex on the short arm of

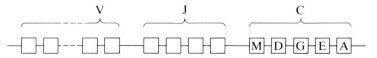

Fig. 11.3 Diagram of the immunoglobulin heavy chain gene

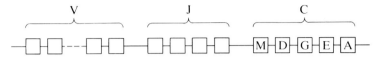

Lymphoid stem cell

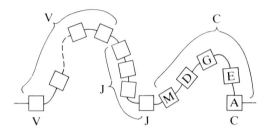

Somatic recombination

Mature lymphocyte
heavy chain gene

Fig. 11.4 Diagram of VJC switching to produce antibodies with different specificities from the same gene complex

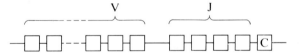

Fig. 11.5 Diagram of the immunoglobulin light chain gene

chromosome 6 but C3 is on chromosome 19 and C8 on chromosome 1.

INHERITED IMMUNODEFICIENCY

Inherited defects may occur in either or both components of the immune response. The symptoms and signs depend upon the residual defence mechanisms but generally include early onset of

Table 11.2 Inherited immunodeficiency

Disease	Mode(s) of inheritance
Severe combined immunodeficiency	AR, XR
Agammaglobulinaemia	XR, AR
Wiskott-Aldrich syndrome	XR
Chronic granulomatous disease	XR, AR
Chediak-Higashi syndrome	AR
Ataxia telangiectasia	AR
DiGeorge syndrome	? chromosomal deletion

undue susceptibility to infections and failure to thrive. As outlined in Table 11.2 these conditions although rare are heterogeneous.

Severe combined immunodeficiency (SCID)

Clinical features
Either sex is affected with onset of failure to thrive and recurrent infections during the first few months of life. As both cellular and humoral responses are absent, viral, fungal and bacterial infections occur. Without a marrow transplant death occurs in infancy. The lymphocyte count is reduced and there is a marked reduction in immunoglobulin levels. No ABO isoagglutinins are present.

Genetic Aspects
SCID shows genetic heterogeneity. Thus whilst the majority of patients are autosomal recessive homozygotes, at least one X- linked recessive form is known. In 20% of patients reduced levels of adenosine deaminase (locus on chromosome 20) are found. In these families the level of adenosine deaminase in amniotic fluid cells can be used for prenatal diagnosis but in the remainder fetal blood sampling to show absence of T and B lymphocyte surface markers is necessary.

The overall incidence of SCID is 2 per million births with a maximal prevalence in Apache Indians.

X-linked agammaglobulinemia

Clinical features
Mostly males are affected with onset around 6 months of age as maternal levels of immunoglobulin G subside. The cellular response is normal and so recurrent bacterial infections especially of the respiratory tract, sinuses and intestines occur. Without therapy stunted growth and arthritis supervene. The serum immunoglobulins are reduced, no ABO isoagglutinins are present and there is

failure to respond to injected antigens. Immunoglobulin replacement is necessary for life.

Genetic aspects
Carrier detection is not possible and prenatal diagnosis requires fetal blood sampling of each male pregnancy to demonstrate absence of B lymphocytes. A similar condition may be inherited as an autosomal recessive trait.

DiGeorge syndrome

Clinical features
Onset is in the neonatal period with hypocalcaemic seizures followed by failure to thrive and recurrent viral and fungal infections. Aortic arch anomalies may be associated and some children show dysmorphic features such as hypertelorism, downslanting palpebral fissures and a fish-like mouth. Immunoglobulins are normal but chest X-ray shows an absent thymus and cell mediated immunity is absent. Despite attempted therapy with a thymus transplant most die in infancy.

Genetic aspects
DiGeorge syndrome is usually a sporadic malformation with aplasia of the third and fourth pharyngeal pouch and fourth branchial arch derivatives. It may be heterogeneous as some patients show an interstitial microdeletion of chromosome 22.

BLOOD GROUPS

Blood groups are determined by antigenic surface proteins on red cells. So far about 400 blood group antigens have been described; some of the best known are indicated in the Table 11.3.

Table 11.3 Examples of human blood groups

Blood group	Chromosomal location
ABO	9
Rhesus	1
Kell	?
Duffy	1
Kidd	2
Lutheran	19
Lewis	19
P	?
MNS	4

ABO BLOOD GROUP

There are four major ABO phenotypes, O, A, B and AB which are determined by the reaction of an individual's red cells with specific anti-A and anti-B antibodies (Table 11.4). The ABO antigens are also present on most other body cells including white blood cells and platelets. Group A individuals possess antigen A on their red cells, group B possess antigen B, group AB have both and group O neither. Group A individuals have IgM anti-B (isoagglutinins) in their serum and group B have anti-A.

The ABO gene is near the tip of the long arm of chromosome 9 and three alleles are identified O, A, and B. There are thus six possible genotypes OO, AA, AO, BB, BO and AB. A and B are inherited as codominant traits with O recessive to both (Table 11.5). It is not possible on blood grouping to distinguish AA from AO or BB from BO. This may however be possible from pedigree

Table 11.4 ABO blood group phenotypes

Red cell phenotype	Reaction with specific antisera	
	Anti-A	Anti-B
O	-	-
A	+	-
B	-	+
AB	+	+
	+ Agglutination − no agglutination	

Table 11.5 Genotypes and phenotypes at the ABO locus

Genotype	Phenotype	UK frequency	Red cell antigens	Serum antibodies
OO	O	0.46	neither	Anti-A and Anti-B
AA AO	A	0.42	A	Anti-B
BB BO	B	0.09	B	Anti-A
AB	AB	0.03	A,B	neither

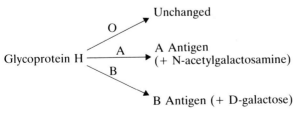

Fig. 11.6 Biosynthesis of ABO blood group substances

information. The ABO alleles determine the activity of specific sugar transferases. Hence the A allele adds N- acetylgalactosamine to the precursor glycoprotein (H substance) whereas the B allele adds D-galactose. In the presence of the O allele the H substance is unchanged (Fig.11.6).

Most persons also secrete the ABO blood group substances into the body fluids such as saliva, sweat, plasma and semen. The ability to do this is determined by the secretor locus on the short arm of chromosome 19. At this locus two alleles Se and se are found and so three genotypes are possible SeSe, Sese and sese (Table 11.6). As only the recessive homozygotes, sese, are unable to secrete their blood group substances into body fluids, the secretor status is determined as an autosomal dominant trait.

Table 11.6 Genotypes and corresponding phenotypes at the secretor locus

Genotype	Phenotype	Frequency	
Se Se	secretor	0.29	⎫
Se se	secretor	0.49	⎬ 0.78
se se	non-secretor	0.22	⎭

Rhesus Blood Group System

There are two main Rhesus phenotypes Rhesus positive and Rhesus negative which are determined by the reaction of an individual's red cells with anti-Rh antibody. Rhesus positive persons possess the Rh antigen on their red cells and other tissues whilst Rhesus negative persons do not.

The Rhesus gene complex is on chromosome 1 with two alleles at each of three closely linked loci C, c; E, e; and D or not D (termed d). Rhesus positive persons are heterozygous or homozygous for the D allele (Table 11.7). In Western European populations 85% are Rhesus postive but the frequency varies in other populations. Nearly all Orientals and North Amerindians are Rhesus positive whereas 30% of Basques are Rhesus negative.

Table 11.7 Rhesus genotypes and phenotypes

Genotype	Shortened notation	Frequency	Rhesus phenotype
cde/cde	rr	0.15	negative
CDe/cde	R_1r	0.32	⎫
CDe/CDe	R_1R_1	0.17	⎮
cDE/cde	R_2r	0.13	⎬ positive
CDe/cDE	R_1R_2	0.14	⎮
cDE/cDE	R_2R_2	0.04	⎭
other rare genotypes		0.05	most positive

Xg Blood Group

There are two Xg phenotypes Xg(a+) and Xg(a-) which are determined by the reaction of an individuals red blood cells with an anti-Xg antibody. The gene for the Xg antigen is near the tip of the short arm of the X chromosome. Males with Xg^aY are Xg(a+) and males with XgY are Xg(a-). The trait is dominant and thus females with Xg^aXg and Xg^aXg^a both test as Xg(a+) (Table 11.8, Chapter 7).

Table 11.8 Xg blood group genotypes and phenotypes

	Genotype	Phenotype	Frequency
Males	Xg^aY	Xg(a+)	0.67
	XgY	Xg(a-)	0.33
Females	Xg^aXg^a	Xg(a+)	0.45
	Xg^aXg	Xg (a+)	0.44
	XgXg	Xg(a-)	0.11

Other Blood Groups

Many other blood groups are known but these are in general of importance only for paternity testing, linkage analysis and research.

HAEMOLYTIC DISEASE OF THE NEWBORN

Haemolytic disease of the newborn is an acquired haemolytic anaemia due to the transplacental passage of maternal immunoglobulin. There are two main causes: ABO incompatibility and Rhesus incompatibility. ABO incompatibility between mother and fetus is relatively frequent but as anti-A and anti-B are predominantly IgM which cannot cross the placenta the clinical disease tends to be mild. In contrast haemolytic disease due to Rhesus incompatibility, although less frequent, is usually severe as anti-D antibodies are IgG which can freely cross the placenta.

Normally during pregnancy small amounts of fetal blood reach the maternal circulation. If mother and fetus are both Rhesus negative or both Rhesus positive then this is of no significance. However, when the mother is Rhesus negative and the fetus is Rhesus positive the fetal cells may stimulate the formation of maternal anti-Rh (D) antibody. Once a Rhesus negative woman has mounted such an immune response she is said to be sensitized. The small transplacental bleeds during pregnancy may be sufficient to cause sensitization but a more significant bleed occurs at the time of delivery and this is the commonest time for sensitization to occur. Sensitization is more likely if the mother and fetus are ABO compatible as this allows the fetal cells to persist in the maternal circulation thus increasing the immune stimulus. A Rhesus negative woman can also be sensitized by a Rhesus positive blood transfusion, by an abortion (either spontaneous or therapeutic) or occasionally by an amniocentesis.

If a sensitised woman has another Rhesus positive fetus maternal anti-Rh will cross the placenta and combine with the Rhesus positive fetal cells. This results in a shortened red cell survival with an increased need for production. Marrow hyperplasia and enlargement of the liver and spleen occur. Severe anaemia produces cardiac failure with generalised oedema (hydrops fetalis) and death may ensue despite intrauterine blood transfusion. In utero the excessive production of unconjugated bilirubin from lysed cells is excreted by the placenta. After birth serum bilirubin levels rapidly rise and can produce brain damage (kernicterus) unless treated by repeated exchange transfusion.

Prevention

In Caucasian populations haemolytic disease of the newborn formerly affected 1% of all births. Typically the first-born was unaffected as this pregnancy caused sensitization. The disease severity then increased with successive pregnancies until intrauterine death became inevitable.

A means of prevention was introduced in 1970. Anti-Rh gammaglobulin is administered to each Rhesus negative non-sensitised woman who delivers a Rhesus positive child. This antibody removes the fetal cells before they can cause sensitization. Anti-Rh antibodies are also given after a Rhesus negative woman has an abortion.

With this technique the incidence of Rhesus haemolytic disease of the newborn has fallen as reflected by its contribution to the stillbirth rate which was 0.52 per 1000 total births in 1968 but only 0.16 per 1000 in 1975. Some women still become sensitized by small bleeds during pregnancy and unfortunately some women at risk are not given anti-Rh gammaglobulin.

PATERNITY TESTING

Non-paternity has medicolegal importance and can also be a problem in genetic counselling.

A series of polymorphic markers including blood groups, tissue type and serum enzymes are studied and paternity can be excluded if:

1. The child possesses a blood group marker not present in either parent.

2.The child is without a marker which is homozygous in the putative father.

3.The child is homozygous for a marker that the putative father does not possess.

With a combination of markers paternity can be excluded in at least 95% of cases. However, paternity can never be proven only excluded.

HLA COMPLEX

The human leucocyte antigen (HLA) complex is the major histocompatibility complex in man. The phenotype is derived from the different combinations of alleles at four adjacent loci A, B, C, and D on the short arm of chromosome 6 (Fig. 11.7). A and B phenotypes are determined by typing with sera from subjects sensitised by prior transfusion; the weaker C system may also be determined by serological methods. Antigens at the major lymphocyte activating locus HLA-D may be identified by looking for non-reactivity in a mixed lymphocyte culture reaction against a selection of homozygous D locus cells. In routine clinical work , however, only HLA A and B loci are determined and the antigenic discrepancy at the D locus is detected with the mixed lymphocyte culture technique.

As these loci are closely linked they tend to be inherited as a unit called a haplotype. A large number of alternative alleles are already known at each of the four HLA loci and doubtless more await discovery. This is the most polymorphic gene cluster so far known in man. This diversity reflects our genetic individuality and renders

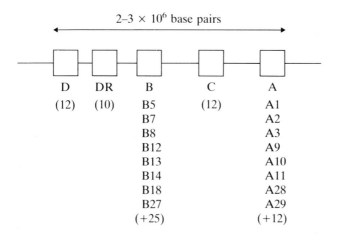

( ) = Provisionally identified alleles

Fig. 11.7 Diagram of the HLA complex

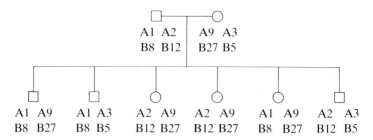

Fig. 11.8 Pedigree showing HLA haplotypes

it most unlikely that unrelated individuals will have identical HLA haplotypes. In a family however there will be a one in four chance that sibs will have identical HLA haplotypes (Fig.11.8).

The HLA gene products are transmembrane proteins associated with beta-2 microglobulin. These antigens are present on all nucleated cells except some trophoblast cells, sperm and choriocarcinoma. The antisera to these antigens are derived from patients who have been sensitized by pregnancy, transfusion or other tissue transplantation.

HLA Disease Associations

Certain HLA antigens show striking associations with certain diseases. The best known of these associations is ankylosing spondylitis with HLA B27. Whereas only 7% of the general population have B27 the frequency in patients with ankylosing spondylitis is 90%. Thus whilst not everyone with B27 will develop ankylosing spondylitis the relative risk is markedly increased. In this context the relative risk is defined as the ratio of the risk of developing the disease in those with the antigen to those without the antigen.

Other less strong associations are known and some diseases are associated with more than one HLA antigen (Table 11.9).

Table 11.9 HLA disease associations

Disease	HLA antigen	Relative risk
Ankylosing spondylitis	B27	90
Reiter's syndrome	B27	40
Multiple sclerosis	Dr2	4
Thyrotoxicosis	B8	11
Myastenia gravis	Dr3, B8	3
Psoriasis	Dr7	43
Insulin dependent	Dr4	6
diabetes mellitus	Dr3	3
	Dr3 and Dr4	33

The significance of these associations is not known. Perhaps the host with the antigen is rendered more susceptible to the environmental agent which causes the disease but the mechanism is obscure.

Linkage Disequilibrium

The alleles at two linked loci are in equilibrium if the frequency of the combination of any two alleles equals the product of the individual allele frequencies. Thus the frequency of the haplotype A3 B27 is 0.021 which equals the product of the individual frequencies (0.31 and 0.07). Within the HLA complex most of the alleles are in equilibrium but there are some important exceptions. A1 has a frequency of 0.17 and B8 of 0.11. Thus the expected frequency of A1 B8 is 0.17 multiplied by 0.11 or 0.019. However, the observed freqeuncy is increased four-fold at 0.09. The combinations A3 B7 and A1 B17 are also found at an increased frequency. When loci like this are not in equilibrium they are said to show linkage disequilibrium.

The cause of linkage disequilibrium is unknown but presumably the combination of A1 and B8 has some selective advantage over either allele alone. The HLA disease associations may similarly be based on linkage disequilibrium to a neighbouring gene, which is perhaps involved in the immune response.

TRANSPLANTATION

Tissue of one individual (the donor) may be transferred to another (the recipient). This transfer is known as a transplant and is further classified according to the relationship of the donor and recipient (Table 11.10).

The transplanted tissue in autografts and isografts is genetically identical to the recipient and so rejection by cell mediated immunity is not a problem. In contrast grafts between different species are always rejected and those between different members of the same species will usually be rejected in the absence of tissue matching and immune suppression therapy.

Table 11.10 Types of tissue transplant

Type	Donor-recipient relationship
Autograft	self
Isograft	identical twins
Allograft	same species
Xenograft	different species

Blood Transfusion

Blood transfusion is the commonest type of tissue transplant. Prior to transfusion the ABO and Rhesus types of donor and recipient are determined and a screen for atypical antibodies is made. Donor cells are also mixed with the recipient's serum in vitro (cross-match).

Generally blood of the same ABO group is given where possible but in an emergency it is possible to give a different type as indicated in the Table 11.11. An ABO antibody present in the plasma of the recipient will cause agglutination of donor cells if they have the corresponding antigen. The donor antibodies are not important as they are rapidly diluted in the recipients circulation.

Table 11.11

		Blood group of donor			
		O	A	B	AB
	O	-	+	+	+
	A	-	-	+	+
Blood group of recipient	B	-	+	-	+
	AB	-	-	-	—

- no agglutination, + agglutination

Other Tissues

A wide variety of other tissues may be transplanted (Table 11.12). Donor selection is critical as host and donor need to be as antigenically similar as possible. Hence relatives provide the best chance of a match as determined by ABO group and HLA tissue type. Compatibility is also tested directly by mixing immune competent lymphocytes from both host and donor (mixed lymphocyte culture test).

If the match is accurate then rejection will not occur. If the match is poor then rejection will occur despite immunosuppresion.

One other complication is specific to marrow transplantation.

Table 11.12   Tissues which can be transplanted

Blood
Bone marrow
Skin
Bone
Kidney
Heart
Liver
Lungs
Corneae
Fallopian tube and ovary

This transplant contains immune competent cells which can mount a cellular response against antigenically dissimilar host cells (graft-versus-host disease).

The fetus is antigenically different from the mother (an allograft) and yet rejection does not occur. The absence of HLA antigens from the outer layer of placental cells may be involved in this as may the presence of fetal white cells in the maternal circulation.

Chapter 12
Human Molecular Pathology

Gene defects are usually identified by studying the particular gene product (ie. its nature or amount) or by identifying the consequences of the abnormal gene product. Advances in recombinant DNA technology have led to a new approach, direct analysis of the gene itself and this has led to a knowledge of the molecular defect in some genetic diseases. Although most interest has focussed on the single gene disorders, especially the haemoglobinopathies, these techniques are also of value for the investigation of the pathology of chromosomal and multifactorial disorders.

1. SINGLE GENE DISORDERS

HAEMOGLOBINOPATHIES

The haemoglobin molecule has four subunits each with a polypeptide globin chain and haem, an iron containing pigment. The haem portion is alike in all forms of haemoglobin. In the normal adult 98% of the haemoglobin is haemoglobin A (HbA) with HbA_2 making up the total. Each molecule of HbA has two alpha globin chains and two beta globin chains. Each alpha chain has 141 aminoacids whereas each beta chain has 146. These chains have a similar but not identical aminoacid sequence both to each other and to myoglobin.

Haemoglobin A_2 has two alpha globin chains per molecule but instead of beta globin chains it has two delta globin chains. There is marked similarity in the structure of delta globin and beta globin. During fetal and early postnatal life the predominant haemoglobin is HbF. This has two alpha globin chains per molecule together with two gamma chains which are 146 aminoacids long. Two types of gamma chains are found which differ only at the 136th aminoacid. In one type it is glycine whereas in the other it is alanine. During early embryonic development other less well characterised types of haemoglobin are present (Fig. 12.1, Table 12.1). Alpha globin chains are present in all types of haemoglobin from the fetus

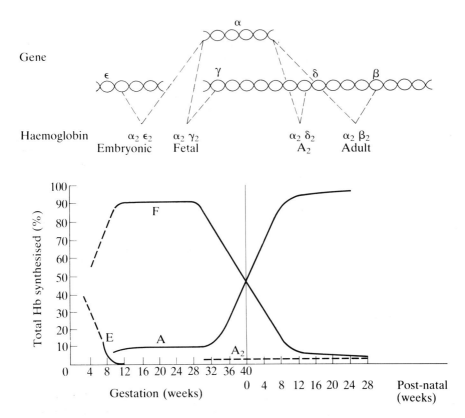

Fig. 12.1 Changes in the predominant type of haemoglobin with age

Table 12.1 Types of human haemoglobin

Type	Structure	Comment
HbA	$\alpha_2\beta_2$	98% of adult haemoglobin
HbA2	$\alpha_2\delta_2$	2% of adult haemoglobin
HbF	$\alpha_2\gamma_2$	major fetal haemoglobin
Hb Gower I	$\zeta_2\varepsilon_2$	
Hb Gower II	$\alpha_2\varepsilon_2$	embryonic haemoglobins
Hb Portland	$\zeta_2\gamma_2$	

onwards whereas the non-alpha chain varies. This accounts for the differences in severity and timing of onset of diseases which interfere with alpha or non-alpha chain production.

The alpha chains are coded by the alpha gene locus on the short arm of chromosome 16. Each chromosome 16 has two adjacent copies of this gene and hence there are four alpha genes in each diploid chromosome set. The non-alpha chain genes lie on the short arm of chromosome 11. Moving from the centromere outwards the order is epsilon, gamma (glycine), gamma (alanine), delta then beta. There is only a single copy of each of these loci. The alpha and

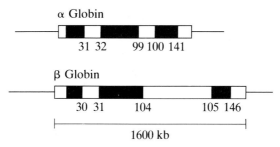

α Globin

31 32   99 100 141

β Globin

30 31   104   105 146

1600 kb

Fig. 12.2 Alpha and beta globin genes. The coding regions (exons) are shaded and the corresponding aminoacids numbered

Table 12.2 Examples of haemoglobinopathy

Type	Molecular defect	Phenotype
HbS	Point mutation, β_6 GLU $\rightarrow$ VAL	Sickle cell disease
HbC	Point mutation, β_6 GLU $\rightarrow$ LYS	
HbE	Point mutation, β_{26} GLU $\rightarrow$ LYS	
HbM Boston	Point mutation, β_{58} HIS $\rightarrow$ TYR	methemoglobinaemia
HbM Saskatoon	Point mutation, β_{63} HIS $\rightarrow$ TYR	methemoglobinaemia
Hb Constant Spring	Point mutation, α_{141} STOP $\rightarrow$ GLN	alpha thalassaemia
Hb Wayne	Deletion at α_{139} $\rightarrow$ frameshift	

beta globin genes have been fully sequenced and as indicated in Fig. 12.2 each has three coding segments (exons) with two intervening sequences (introns).

More than 350 abnormal haemoglobins have been described. Most of these variants result from point mutations in the structural genes which code for the aminoacid sequence of the globin chain of the molecule (ie. multiple alleles). Many do not interfere with the function of the haemoglobin and are thus asymptomatic. Others at more critical functional sites produce anemia, cyanosis or methaemoglobinemia. Table 12.2 indicates some of the commoner clinically important haemoglobinopathies.

Sickle cell disease

Sickle cell disease is caused by a point mutation in the beta globin gene. This alters the corresponding mRNA codon from GA(A or G) which codes for glutamic acid to GU(A or G) which codes for valine. The globin chain produced by the mutant gene is called haemoglobin S and only differs from HbA in having valine instead of glutamic acid at the sixth position from the N- terminal end (Fig. 12.3). This difference alters the electrophoretic mobility (Fig. 12.4) and at reduced oxygen tension HbS tends to aggregate into rod-like masses which distort the red cells into sickle shapes - hence the name of the disease (Fig. 12.5).

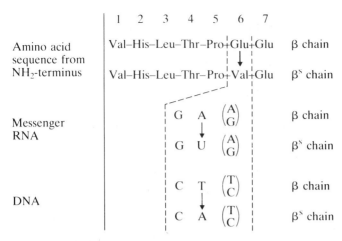

Fig. 12.3 Molecular pathology of haemoglobin S

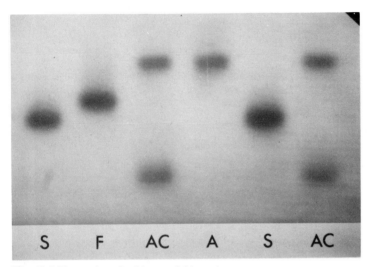

Fig. 12.4 Electrophoresis of haemoglobin variants

An individual with one normal beta globin gene and one sickle cell disease gene is clinically normal. Occasionally, especially if exposed to a reduced oxygen pressure for example at high altitude or during general anesthesia, these heteroygotes have renal or splenic infarcts. This heterozygous state is known as sickle cell trait and can be detected by exposing the red cells in vitro to a very low oxygen tension and by electrophoresis which shows 30-45% HbS with HbA making up the total.

Homozygotes for this mutant beta globin gene have sickle cell disease. The fetus is unaffected as production of haemoglobin F is unimpaired. During infancy, however, as the beta globin genes become active cells with HbS are produced. A severe chronic haemolytic anaemia ensues as these distorted red cells have a

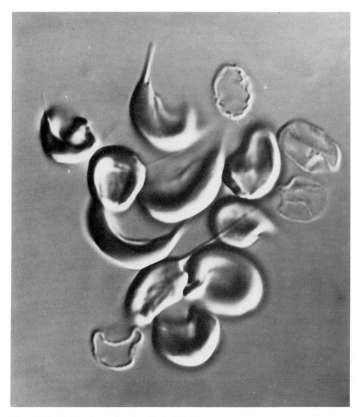

Fig. 12.5 Sickle-shaped red cells in HbS homozygote

reduced survival time. Distorted red cells may also occlude vessels with recurrent infarctions especially of the lungs, bones and spleen. The marrow shows compensatory hyperplasia. The spleen is enlarged in early childhoood but usually becomes impalpable later when subject to recurrent infarction (autosplenectomy). Affected persons are also prone to pneumococcal infections and to Salmonella osteomyelitis. The blood film shows anaemia with distorted red cells (Fig.12.5). Haemoglobin electrophoresis shows mainly HbS, with some HbA_2 and some persistence of HbF (5-15%). Despite supportive care affected individuals have a short-ened lifespan although they may reach adulthood and reproduce.

The disease has a characteristic geographical distribution which mirrors the distribution of Falciparum malaria (Chapter 10). Table 12.3 indicates the frequency of the disease and the carrier state in different ethnic groups. It is estimated that 100,000 homozygotes are born each year in Africa, 1,500 in the USA, 700 in the Carribean and 140 in the UK.

For heterozygous parents there is on average a 1 in 4 risk that each child will be homozygous. Prenatal diagnosis formerly required fetal blood sampling and determination of beta globin chain

Table 12.3 Frequency sickle cell trait and disease

Ethnic group	Frequency of sickle cell disease	Carrier frequency
African negroes	1/40	1/3
US negroes	1/400	1/10

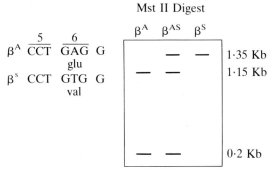

β^A $\overline{\dfrac{5}{\text{CCT}} \dfrac{6}{\text{GAG}}}$ G
glu

β^S CCT GTG G
val

Mst II Digest

β^A β^{AS} β^S

1·35 Kb
1·15 Kb

0·2 Kb

Fig. 12.6 Restriction fragment patterns in normal homozygote (β^A), heterozygote (β^{AS}) and sickle cell homozygote (β^S) (Mst II digestion)

synthesis but is now possible by direct DNA analysis after amniocentesis or chorion biopsy. The point mutation which causes sickle cell disease also removes a recognition site for a restriction endonuclease (Mst II). When DNA from a normal person is digested with this enzyme the beta globin gene yields fragments of 1150 and 200 basepairs; a carrier produces fragments 1150 bp, 200bp and 1350 bp in length; and a homozygous affected person yields only the 1350bp fragments (Fig. 12.6).

Other globin point mutations

In addition to the aminoacid substitution as seen in HbS, point mutations may lead to premature chain termination, delayed chain termination or shift of the codon reading frame.

Haemoglobin E is found mainly in Orientals and reaches a frequency of up to 50% in parts of Thailand. It is due to a beta globin point mutation which substitutes lysine for glutamic acid at position 26. This point mutation results in the loss of recognition sites for two enzymes (HphI and MnII) and alters the haemoglobin electrophoretic mobility. Heterozygotes are asymptomatic and homozygotes have only mild anaemia.

Haemoglobin M results from aminoacid substitutions close to the attached haem. These keep the haem in the ferric state and so result in methemoglobinemia.

Haemoglobin Constant Spring is caused by a point mutation at the normal alpha chain terminator. Messenger RNA thus continues to be transcribed until the next stop codon is reached. The final product is an alpha chain with 172 aminoacids (normal 141). This abnormal alpha chain has a slow rate of synthesis and so affected individuals have a disease similar to mild alpha thalassemia.

Haemoglobin Wayne is due to an alpha chain mutation with loss of the third nucleotide of the 139th codon. This causes a frame shift and results in an abnormal alpha chain.

Alpha thalassaemia

The alpha thalassaemias are due to a reduced rate of alpha globin synthesis. Reduction in synthesis is most often caused by deletion of one or more of the alpha globin genes. Normally four copies of the alpha globin genes are present, two per short arm of each chromosome 16 ($\alpha\alpha/\alpha\alpha$). Deletion of all four genes (--/--) interferes wih haemoglobin production from early fetal life. Anaemia is profound and causes heart failure with generalised oedema (hydrops fetalis). Intrauterine death is usual and maternal pre-eclampsia is severe. The only haemoglobin present is abnormal with four gamma chains (Hb Barts).

If only one alpha gene is present ($-\alpha/--$) the individual has HbH disease. Affected persons have a chronic haemolytic anaemia with reduction in the mean red cell haemoglobin and volume. Lifespan is, however, normal. Haemoglobin H which is composed of four beta chains accounts for 5-30% of the total haemoglobin with HbA as the remainder. One parent of a patient will have two missing alpha genes ($--/\alpha\alpha$) and the other will have one missing alpha gene ($-\alpha/\alpha\alpha$). The absence of two alpha genes ($--/\alpha\alpha$) is called alpha thal 1 trait and is asymptomatic. The blood count will show a mild hypochromic microcytic anaemia with 5-6% HbH in the cord blood which disappears by six months of age. The absence of only one alpha gene ($-\alpha/\alpha\alpha$) is called alpha thal 2 trait and produces no clinical or haematological features. A person who is homozygous for the alpha thal 2 trait ($-\alpha/-\alpha$) will clinically be indistinguishable from the alpha thal 1 heterozygote (Fig. 12.7).

Population carrier surveys are more difficult than for beta thalassaemia and so data on the frequency of alpha thalassaemia genotypes is more limited (Table 12.4).

Molecular studies have revealed that most cases of alpha thalassaemia are due to gene deletion although of variable extent. These deletions are believed to arise by unequal crossing over at meiosis which is especially likely wherever linked genes of similar structure occur thus favouring misalignment (Fig. 5.10).

For parents who both have alpha thal 1 trait ($--/\alpha\alpha$) there is a 1 in 4 risk of producing a child who has no alpha globin genes (hydrops fetalis, alpha thal 1 homozygote). Prenatal diagnosis is possible by direct DNA analysis after aminocentesis or chorion biopsy.

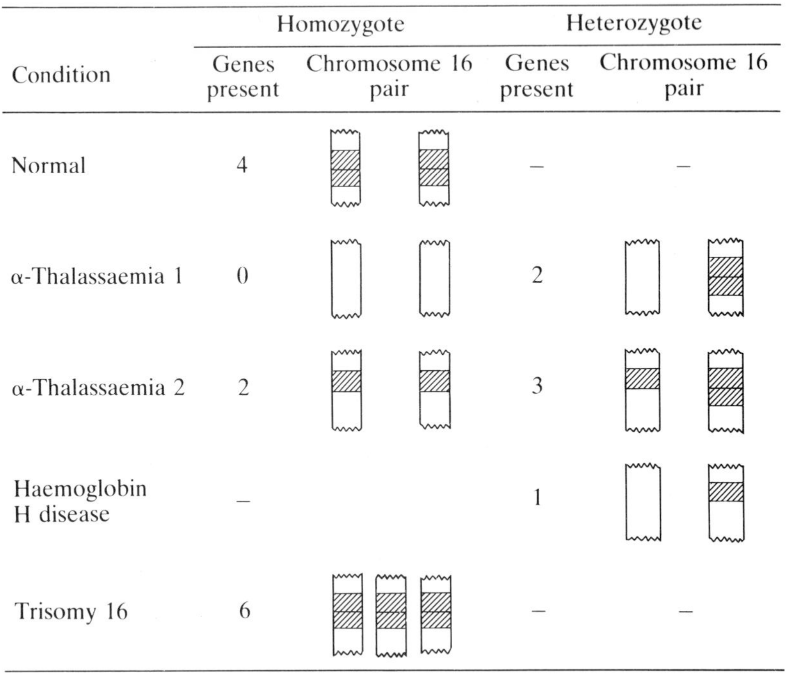

Condition	Homozygote		Heterozygote	
	Genes present	Chromosome 16 pair	Genes present	Chromosome 16 pair
Normal	4		–	–
α-Thalassaemia 1	0		2	
α-Thalassaemia 2	2		3	
Haemoglobin H disease	–		1	
Trisomy 16	6		–	–

Fig. 12.7 Alpha globin gene dosage in alpha thalassaemia

Table 12.4 Estimates of alpha thalassaemia genotypes in various ethnic groups

Ethnic group	Genotype	Frequency
Thais	-α/αα	5-8%
	--/αα	6%
	--/--	0.4%
Negroes	-α/αα	25%
	--/αα	rare

Digestion of the DNA with a restriction enzyme (EcoRI) yields fragments of different lengths from the alpha globin region in proportion to the number of remaining genes (Fig. 12.8).

Beta thalassaemia

The beta thalassaemias are due to a reduced (β^+) or absent (β°) beta chain synthesis. A variety of molecular defects may interfere with normal beta gene expression but deletion as seen in alpha thalassaemia is infrequent. Table 12.5 indicates some of the known mutations in the beta globin gene (multiple alleles) which cause beta thalassaemia. Homozygosity (identical alleles) or any combination of two of these mutant alleles (a genetic compound) can produce beta thalassemia. This genetic heterogeneity accounts for the observed range in clinical severity.

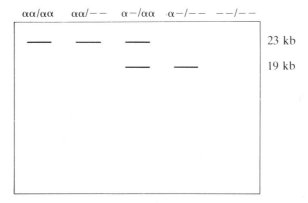

Fig. 12.8 Restriction fragment patterns in normal and various alpha thalassaemia genotypes (EcoRI digestion)

Table 12.5 Molecular pathology of beta thalassaemia

Molecular mechanism	Examples
1. Point mutation	
(i) premature chain terminaton	
	β_{17} LYS → STOP (Chinese)
	β_{39} GLN → STOP (Italian)
(ii) frameshift	two base deletion at β_8 (Turkish)
(iii) abnormal mRNA splicing	
	G → A at intron 1/ exon 2 junction (Cypriots) - loss of site
	G → A at exon 2/intron 2 junction (Italians) - loss of site
	β_{26} GAG → AAG creates new site in exon 1
2. Partial deletion	600bp (Indian)
3. Insertion	
4. Defective mRNA	(Kurdish Jews)

Beta thalassaemia heterozygotes with one mutant and one normal beta globin gene are asymptomatic but have a mild microcytic hypochromic anaemia with target cells. This is similar to the blood film of iron deficiency anaemia but can be distinguished by the presence of increased amounts of HbF and HbA_2.

The homozygote with two mutant beta globin genes has unimpaired production of fetal haemoglobin but with the changeover to HbA in infancy a severe anemia ensues. Marrow hyperplasia occurs in an attempt to compensate for faulty erythropoiesis and results in bony expansion with skull bossing and hepatosplenomegaly. Severe hypochromia, microcytosis with target cells and increased HbF and HbA_2 occur. Repeated blood transfusions are required and may lead to chronic iron overload

Table 12.6 Estimates of beta thalassaemia heterozygote frequency in various ethnic groups

Ethnic group	Carrier frequency
Cypriots	1/6
Greeks	1/14
Italians	1/10 - 1/50
Indians	1/6 - 1/50
Turkish	1/50
Thais	1/10 - 1/50
Chinese	1/50
US Negroes	1/70

unless treated with the iron-chelating agent desferrioxamine. Despite these supportive measures lifespan is reduced.

The heterozygous state is believed to confer a selective advantage against Falciparum malaria (Chapter 10) and so the mutant beta globin genes have reached high frequencies in certain ethnic groups (Table 12.6). It is estimated that worldwide about 100,000 severely affected homozygotes are born each year.

If parents are both heterozygotes there is on average a 1 chance in 4 that each child will be homozygous affected. Prenatal diagnosis is possible and in 60% of families can be performed by DNA analysis. The remainder need fetal blood sampling with demonstration of the reduced rate of beta globin chain synthesis. As beta thalassaemia is so genetically heterogeneous at the molecular level the demonstration of loss of a recognition site (as in HbS) or of a deletion (as in alpha thalassaemia) is not applicable. Recognition sites for several enzymes on either side of the beta globin locus are compared for each chromosome 11 in the parents and the affected children. Thus in Fig. 12.9 the presence (+) of a recognition site for a particular enzyme identifies the chromosome 11 in each parent which carries the mutant allele. If the fetus inherits each chromosome 11 with this marker then both tightly linked mutant beta genes will also have been inherited and the fetus would be affected. In Fig. 12.9 the fetus is a heterozygote.

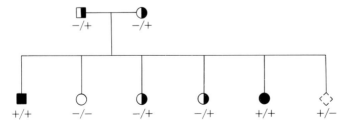

Fig. 12.9 Prenatal diagnosis of beta thalassaemia using the pattern of enzyme recognition sites in the beta globin region to identify which homologues of chromosome 11 carry the mutant genes

Hereditary persistence of HbF

Normally HbF is only produced in significant amounts during fetal life. The regulation of haemoglobin type synthesis is unknown but if a deletion occurs between the gamma and beta genes on chromosome 11 then this results in abnormal persistence of HbF. Heterozygotes have 20-30% HbF and in homozygotes 100% of the haemoglobin in adult life is HbF. These individuals have no symptoms and the coexistence of this trait with sickle cell disease (a beta globin defect) ameliorates the clinical features.

Haemoglobinopathy compounds

There is thus a wide variety of molecular pathology in the genes involved in haemoglobin production. Many of these mutant alleles at the alpha or beta loci reach high frequencies in certain ethnic groups and not uncommonly different mutant alleles may coexist in the same individual. The combination may be advantageous (eg. HbF persistence and HbS) or disadvantageous (eg. HbC and HbS).

OTHER SINGLE GENE DISORDERS

Most human single gene disorders are likely to be as heterogeneous at the molecular level as the haemoglobinopathies but as yet much less is known about their molecular pathology.

Inherited disorders of collagen

Collagen is the most important structural protein in the body. It forms the framework of all tissues and so far 8 types have been identified (Table 12.7). Type I collagen is the most abundant and is (like the other types) composed of three alpha chains twined about each other. In type I collagen two alpha 1(I) and one alpha 2(I) chains are found. These differ in aminoacid composition from each other and from the other collagen chains. The gene for the alpha 1 chain is on chromosome 17 and that for alpha 2 is on chromosome 7.

Table 12.7 Types of collagen

Type	Distribution	Composition	Gene locus
I	bone, skin, tendon, vessels, uterus	two alpha 1(I) one alpha 2(I)	17q21-22 7q21-22
II	cartilage, cornea	three alpha 1(II)	?12p
III	skin, vessels, lung uterus, gut	three alpha 1(III)	?7q
IV-VIII basement membranes			

Table 12.8 Examples of molecular pathology in inherited disorders of collagen

Phenotype	Molecular pathology
Osteogenesis imperfecta	partial deletion of alpha 1(I)
	partial deletion of alpha 2(I)
	point mutation in C propeptide
	non-functioning allele of alpha 2(I)
Marfan syndrome	insertion of 20 aminoacids in alpha 2(I) chain

These genes are very large and complicated; each has about 50 coding segments (exons). Collagen also undergoes extensive post-translational processing and thus there are many points at which collagen synthesis could be disrupted.

Several different mutations in the collagen genes have already been discovered in inherited collagen disorders and undoubtedly further heterogeneity of molecular pathology will be found (Table 12.8).

Growth hormone deficiency

The gene for growth hormone is close to the collagen alpha 1 gene on chromosome 17. Rarely growth hormone deficiency may be inherited as an autosomal recessive trait and in some of these families the growth hormone gene is deleted. The heterozygotes are of normal stature as they produce sufficient hormone from their remaining normal allele.

Haemophilia B (Factor IX deficiency)

In some families with this type of haemophilia the gene for factor IX (location Xq27) is deleted, in others the gene is present but non-functional.

2. CHROMOSOMAL DISORDERS

HbH and mental retardation

Several children have been identified with mental retardation and haemoglobin H. In this form of alpha thalassaemia there is loss of three of the four alpha genes ($-\alpha/--$). In each case one parent had alpha thal 2 ($-\alpha/\alpha\alpha$) but the spouse had normal alpha genes (α/α). Thus the child had inherited the single alpha gene from one parent and had a new deletion in the alpha globin region on the opposite chromosome. Molecular analysis confirmed the existence of this new deletion which varied in extent between affected individuals.

Although known to be present this deletion was not visible with the light microscope.

This serves to emphasise the resolution gap between pathology at the molecular level and visible chromosomal aberrations. The smallest visible chromosomal loss is about 6000 kilobases but the average gene is only about 2 kilobases in length. These submicroscopic deletions may be important in the aetiology of at least some cases of idiopathic mental retardation or unexplained multiple congenital malformations.

3. MULTIFACTORIAL DISORDERS

CANCER

The first step in many cancers is a gene mutation. This can be in the form of an inherited mutation, as in polyposis coli or xeroderma pigmentosum (Chapter 15), or a somatic mutation generated for example, by irradiation or a carcinogen. It seems that the initial mutation is seldom sufficient to cause neoplasia and that there must be the coincidence of at least one and sometimes more chromosomal changes before the cell loses control. This has been demonstrated for retinoblastoma which seems to result when a retinal cell becomes homozygous for a mutation carried on band 14 of the long arm of chromosome 13. A familial mutation at this site on one chromosome 13 predisposes carrier individuals to this cell specific malignancy. A tumour develops when the normal allele at the retinoblastoma gene locus is lost or is exchanged for the mutant allele by somatic recombination between the two chromosomes 13 (Chapter 3). These events have been shown in retinoblastoma cells by a combination of chromosome analysis and molecular genetic studies. It is still not yet known what the normal product of the retinoblastoma gene is, nor how it leads to neoplasia. It may be that the retinoblastoma gene is an oncogene and similar to those carried by many oncogenic viruses (eg. the Rous sarcoma virus). Oncogenes have been highly conserved in evolution and at least 20 different types occur in the normal human genome. Identical genes have been found in species as distant as man, the earthworm and the fruit fly. These genes may produce important cellular growth factors which can cause cancer if normal regulation is impaired. In the evolution of oncogenic viruses, the ancestral virus incorporated into its own genome an oncogene from its host which it henceforth used to induce tumour formation and so ensure its more successful propagation. Some viruses can create tumours without adopting an alien oncogene, simply by inserting their own promoter DNA sequences in front of a host oncogene which is then activated and leads in some cases to uncontrolled production of the host's cellular growth factor. Presumably a mutation at the host's promoter or regulatory DNA sequences could result in the activation of an intrinsic oncogene, and it is possible that irradiation or carcinogens sometimes act in this way.

Two other ways in which human oncogenes can be activated to form cancer have been discovered recently. In human bladder cancer a point mutation in the host oncogene *H-ras* appears to be sufficient to cause neoplastic transformation. A similar mutation at exactly the same site in the related oncogene *K-ras* has been shown to occur in certain lung cancers. The coincidence is remarkable and seems to point not only to the possibliity of a limited number of oncogenes, but also to specific sites where mutation is critical for tumour formation.

The third way in which human oncogenes can be activated to cause cancer is by the process of chromosomal translocation. In chronic myeloid leukaemia the *c-abl* oncogene is translocated from its normal site at the end of the long arm of chromosome 9 to a site on chromosome 22 (the Philadelphia chromosome) next to the gene for lambda immunoglobulin light chains. It seems that the DNA promoters and related sequences normally involved in the production of lambda light chains, now activate the *c-abl* oncogene instead. Similarly, in Burkitt's lymphoma the characteristic chromosomal translocation between chromosomes 8 and 14 brings the *c-myc* oncogene on chromosome 8 to the site of the immunoglobulin heavy chain gene on chromosome 14. The EB virus and another oncogene *B-lym* are also involved in Burkitt's lymphoma but their relationship with *c-myc* has not yet been clarified.

Once a cancer is initiated it may become progressively more malignant and these changes are often associated with multiple chromosome aberrations and evidence of involvement of other oncogenes. In some malignancies (including lung cancers) the number of copies of a particular oncogene can be amplified by the generation of multiple repeats or in tiny fragments termed double minutes. The effect is to increase the production of the cellular growth factors and thus the rate of progression of the tumour.

OTHER MULTIFACTORIAL DISORDERS

For the vast majority of multifactorial disorders the nature and number of interacting alleles is unknown (Chapter 16). As in the case of cancer molecular studies may help to unravel their aetiology. Gene probes are now available for the insulin gene and for the apoprotein genes; such probes could be used to study families with diabetes mellitus and ischaemic heart disease.

Chapter 13
Genetic Counselling

Genetic counselling is the communication of information and advice about inherited conditions. A person seeking genetic advice is called a consultand.

There are five stages in this communication process:
1. History and Pedigree construction
2. Examination
3. Investigations
4. Counselling
5. Follow-up

1. HISTORY AND PEDIGREE CONSTRUCTION

The affected individual who caused the consultand(s) to seek advice is the proband. Often the proband is a child, but he or she may also be the consultand or a more distant relative. A standard medical history is required for the proband and for any other affected persons in the family.

Next the pedigree is constructed. A standardised set of symbols are used (Fig.13.1). The male line is conventionally placed on the left and all members of the same generation are placed on the same horizontal level. Roman numerals are used for each generation starting with the earliest and arabic numerals are used to indicate each individual within a generation (numbering from the left). Thus in Fig. 13.2 III4 is the proband and his parents who are seeking advice are II5 and II6. When drawing a pedigree it is advisable to start at the bottom of the page with the most recent generation and work upwards. The offspring of each set of parents are given in birth order with the eldest on the left. For each member of the pedigree name and age are included. For an extended family study the full name, age, address and phone number for individuals who need to be contacted should be included. Miscarriages, neonatal deaths and consanguinity are often not mentioned unless specifically asked about.

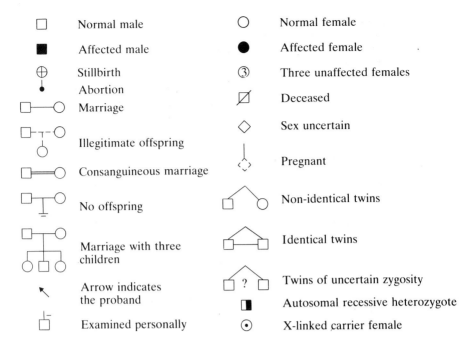

Fig. 13.1 Symbols used for pedigree construction

Normal male   Normal female
Affected male   Affected female
Stillbirth   Three unaffected females
Abortion
Marriage   Deceased
    Sex uncertain
Illegitimate offspring
    Pregnant
Consanguineous marriage
    Non-identical twins
No offspring
    Identical twins
Marriage with three
children   Twins of uncertain zygosity
    Autosomal recessive heterozygote
Arrow indicates
the proband   X-linked carrier female
Examined personally

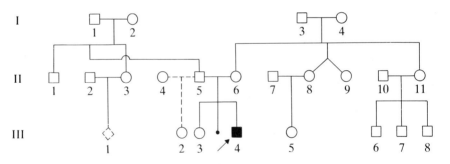

Fig. 13.2 Example of a family pedigree

2. EXAMINATION

A complete physical examination of the proband is desirable. This examination differs from the routine, however, in that there is often a need to accurately describe dysmorphic features. A dysmorphic feature is by definition some characteristic which is outside the range seen in normal individuals. Table 13.1 and Fig. 13.3 indicate some of the terms used for the description of dysmorphic features. In Caucasians a line normally passes through the inner and outer canthi of both eyes and the upper ear attachment lies on or above this line with the head upright. This line divides the adult face into two halves whereas in children the face is divided by the brow line.

Table 13.1 Descriptive terms used in dysmorphology

Term	Meaning
Hypertelorism	Interpupillary distance above expected
Hypotelorism	Interpupillary distance below expected
Telecanthus	Inner canthal distance above expected yet interpupillary distance not increased
Low set ears	Upper border of ear attachment below intercanthal line with head upright
Mongoloid slant	Outer canthi above inner canthi
Antimongoloid slant	Inner canthi above outer canthi
Brushfield spots	Speckled iris ring (20% of normal babies)
Simian crease	Single transverse palmar crease
Epicanthic folds	Skin folds over inner canthi
Brachycephaly	Short antero-posterior skull length
Dolichocephaly	Long antero-posterior skull length
Clinodactyly	Incurved fifth fingers

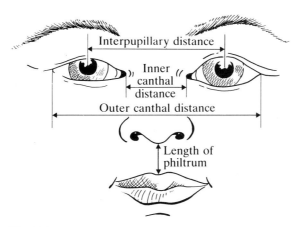

Fig. 13.3 Facial landmarks

The normal facial width is about two-thirds of its height. Clinical impressions can be misleading and so it is desirable to make accurate measurements to confirm features such as widely spaced eyes or disproportionate short stature. Table 13.2 gives some commonly used measurements in this context. The normal ranges of each will vary with age and sex (see standard dysmorphology texts in reference section). Normally each measurement is close to the same percentile. If not then the unusual measurements reflect an abnormality. For example, if height and head circumference are on the 10th percentile but interpupillary distance is on the 90th percentile then relative hypertelorism is present although the actual measurement is within the normal range.

An often neglected aspect of the physical examination is the recording of the finger print pattern (dermatoglyphics). Abnormal

Table 13.2 Standard dysmorphic measurements

Measurement	Comment
Height	
Weight	
Lower segment	Floor to upper border of pubis
Upper segment	Height minus lower segment
Interpupillary distance	Fig. 13.3
Inner canthal distance	Fig. 13.3
Head circumference	Maximum occipitofrontal circumference
Testicular volume	
Ear length	Maximum ear length

Table 13.3 Dermatoglyphic abnormalities

Condtion	Dermatoglyphic findings
Trisomy 18	6-10 arches, Simian crease (30%)
Turner's syndrome	predominance of whorls
47,XXY	excess of arches
5p-	excess of arches, Simian crease (90%)
Trisomy 13	excess of arches, Simian crease (60%)
Trisomy 21	usually all ulnar loops, Simian crease (50%)

Arch   Loop   Whorl

Fig. 13.4 Finger print patterns

dermatoglyphics may provide important clues to the diagnosis (Table 13.3). Three common patterns are found: loop, arch and whorl (Fig. 13.4). Loops are further subdivided into ulnar and radial according to the side of the forearm from which they point. In 4% of the population only a single palmar crease (Simian crease) is present on one hand and in 1% of the normal population single palmar creases are present bilaterally.

In patients with multiple dysmorphic features it is necessary to consider identifiable syndromes. A syndrome (Gk. running together) is the occurrence in the same individual of two or more abnormalities. A few generalisations are possible. Most syndromes have multiple components of which few if any are universal or pathognomonic features. Thus the average patient will not have every feature listed in the textbook. Furthermore some abnormalities are non-specific for example reduced height and high-arched palate may be seen in severe mental retardation of any cause.

3. INVESTIGATIONS

The history and physical examination may permit a confident diagnosis or may indicate the need for further investigation. A wide variety of investigations may be required reflecting the wide spectrum of genetic disease. Indications for chromosomal analysis are given in Table 13.4. Chromosomal abnormalities may produce diverse dysmorphic features and chromosomal analysis should be considered if these are marked especially if accompanied by mental retardation.

There is little point performing chromosomal analysis on patients with single congenital malformations, single gene disorders or with recognisable non-chromosomal syndromes.

Sometimes the affected individuals will have died or be otherwise unavailable for assessment. An attempt should then be made to obtain hospital or other records which might aid definitive diagnosis.

Table 13.4 Indications for chromosomal analysis

Dysmorphic features suggestive of a chromosomal syndrome
Unexplained mental retardation *
Family study of structural chromosomal abnormality
Multiple congenital abnormalities
Unexplained stillbirth
Female with unexplained short stature
Recurrent miscarriages
Primary infertility
Ambiguous sexual development
Leukaemia

---

*Include cultures in folate deficient medium to exclude fragile X

4. COUNSELLING

Accurate diagnosis is fundamental for sound genetic counselling. Hence counselling should never precede the steps involved in diagnosis as outlined above. Both parents should be counselled and adequate time allowed in an appropriate setting. Few couples can be couselled in under 30 minutes and neither the corner of a hospital ward nor a crowded clinic room are adequate. Further it is inappropriate to counsel too soon after recent bereavement or after the initial shock of a serious diagnosis.

Counselling needs to include all aspects of the condition and the depth of explanation should be matched to the educational background of the couple. One might start by outlining the clinical features, complications, natural history, prognosis and therapy (if any) of the condition. Then a simple explanation of the genetic basis of the condition, perhaps with the aid of a diagram, could be given and a recurrence risk calculated for the consultands. It is often useful to compare this recurrence risk against the general population risk for the condition and for other common birth defects (Table 13.5). Generally geneticists consider a risk of more than one in ten as high and or less than one in twenty as low but risks have to be considered in relation to the degree of disability.

Consultands often feel very guilty or stigmatised and it is important to recognise and allay this. Common misconceptions about heredity may also need to be dispelled (Table 13.6).

The reproductive options open to the couple may now be discussed (Table 13.7). In many consultations the couples' fears are unjustified and they can undertake a pregnancy with the reassurance that their risk of genetic disease is no different from other couples in the general population. Where there is an increased risk and especially where the disease burden is significant then other options need to be considered. In this context disease burden is the

Table 13.5 General population risks

Condition	Risk
Spontaneous miscarriage	1 in 6
Perinatal death	1 in 30-100
Neonatal death	1 in 150
Cot death	1 in 400
Major congenital malformation	1 in 33
Serious mental or physical handicap	1 in 50
Adult cancer	1 in 10

Table 13.6 Common misconceptions about heredity

1. Absence of other affected family members means that a disorder is not genetic and vice versa
2. Any condition present at birth must be inherited
3. Upsets, mental and physical, of the mother in pregnancy cause malformations.
4. Genetic diseases are untreatable.
5. If only males or females are affected in the family this indicates sex linkage.
6. A 1 in 4 risk means that the next three children will be unaffected.

Table 13.7 Reproductive options

Proceed to a further pregnancy— with prenatal diagnosis
                                                  — without prenatal diagnosis
Artificial insemination by donor (AID)
In vitro fertilization
Adoption
Contraception — reversible
                      — non-reversible

consultand's perception of the cost, physical, emotional and financial of the disorder. The possibility of prenatal diagnosis for the condition needs to be considered as if available this often encourages a couple to undertake a further pregnancy which otherwise they would be reluctant to risk. Where the couple decide not to undertake a further pregnancy it is necessary for the counsellor to ensure that contraception is adequate and to mention other means of family extension. About 1% of all artificial insemination by donor (AID) is performed for genetic indications such as a husband with an autosomal dominant trait or both parents carriers for an autosomal recessive disease. Although this substantially reduces the risk of an autosomal recessive trait, some risk will remain in proportion to the population carrier frequency. Adoption has been an increasingly difficult alternative in recent years.

Counselling must be non-directive. The aim is to deliver a balanced version of the facts which will permit the consultands to reach their own decision with regard to their reproductive future.

For certain conditions, such as balanced chromosomal rearrangements, autosomal dominant traits and X-linked recessive traits, an extended family study will be required and it is useful to enlist the aid of the consultands in approaching other family members at risk.

5. FOLLOW-UP

Most consultands can be fully counselled at the one sitting but some will require follow-up sessions. Our policy is to follow the counselling with a letter to the consultands which summarises the information given and invites them to return if new questions arise.

WHO NEEDS COUNSELLING?

In the West of Scotland, which has a population of 3 million, we estimate at least 3000 families are in need of genetic counselling each year. We counsel about 1000 of these families each year. Some of the remainder are adequately counselled by other informed and interested health professionals but we suspect that a substantial number still do not receive the information they need.

ETHICAL AND LEGAL ASPECTS

Genetic counselling

In the UK under the Congenital Disabilities (Civil Liability) Act of 1976 legal action can be brought against a person whose breach of duty to parents results in a child being born disabled, abnomal or unhealthy. In both the UK and the USA there was a escalation of litigation concerning genetic disease in the 1970's. Most cases concerned doctor errors of omission and it behoves all doctors who provide genetic advice to ensure the validity and up-to-dateness of that information. Thus failure (whether from ignorance or religious objections) to give correct advice or to refer to someone who would do so on the risk of fetal abnormality in a future pregnancy and the possibility of prenatal diagnosis may constitute medical negligence. The claim would fail if at any time both parents knew and accepted that the child might be abnormal.

Prenatal diagnosis

Prenatal diagnosis with selective termination of pregnancy became a reality in the UK with the Abortion Act of 1967. Under this Act one ground for termination of pregnancy is "..a substantial risk that if the child were born it would suffer from such physical or mental abnormality as to be seriously handicapped". The legal position in other countries varies from total prohibition of abortion for fetal abnormality as in Eire and Canada to relative liberality in the USA and Eastern Europe. Prenatal diagnostic tests need informed consent and parents should be reminded that no single test excludes all known fetal abnormalities and that occasionally the test fails to give a result.

Consanguinity

All human societies in existence prohibit the mating of first degree relatives (incest). Marriage between relatives less close than sibs or parents and offspring is not necessarily outlawed but the dividing line between legal and illegal varies somewhat between countries. Thus in about one half of the United States uncle-niece, aunt-nephew and first cousin matings are forbidden by law. In most African societies consanguineous marriage is not allowed. In contrast, in parts of Japan and India marriage between relatives is encouraged and up to 10% of marriages are consanguineous. The marriage of double first cousins (both sets of grandparents in common) is the closest legal union in the UK.

SPECIAL POINTS IN COUNSELLING

There are several pitfalls for the unwary who practise genetic counselling (Table 13.8). Precision of diagnosis is fundamental to meaningful genetic counselling and most mistakes arise from an inaccurate or incomplete diagnosis. Incomplete knowledge of the

Table 13.8 Pitfalls in genetic counselling

Incorrect or incomplete diagnosis
Genetic heterogeneity
Non-penetrance
Variable expression
Inadequate knowledge of the literature
Previously undescribed disease

literature is especially important for syndromic assessment and in relation to genetic heterogeneity.

In the following sections some general points are made but the counselling details for individual conditions are given in Chapters 14-17.

Chromosomal disorders

The exact recurrence risk varies with the abnormality but for couples at high risk prenatal diagnosis is always an option. Parental karyotypes need not be performed if the child has regular aneuploidy but must be examined if the child has a partial duplication or deficiency. Extended family studies will be required if a parent has a balanced structural rearrangement.

Autosomal dominant traits

The risk to each child of an affected person is on average one in two whereas the risk to the offspring of an unaffected person is negligible provided the disorder has high penetrance. This information needs to be modified accordingly as the penetrance falls. Virtually all dominant traits show variable expression. Where dominant traits have an age-dependent onset such as Huntington's chorea the conditional information that a person is unaffected at a given age can be combined with the pedigree risk using Bayes' theorem (Appendix 2). Extended family studies may be required to counsel all at risk for the trait.

Autosomal recessive traits

Fig. 13.5 shows a family with an autosomal recessive condition. For the carrier parents the risk of recurrence is on average one in four. For each normal sib there will be a two in three chance of being a carrier (not three in four as one possibility namely homozygous affected is excluded). The carrier risk is indicated for other family members. The carrier frequency in the general population needs to be known in order to provide a recurrence risk for relatives other then the parents. For example, the chance that the unaffected sib of

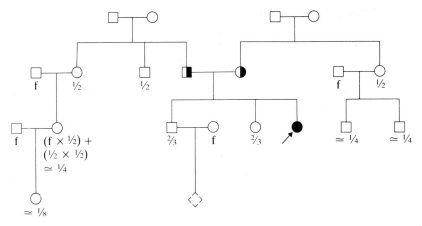

Fig. 13.5 Autosomal recessive trait in a family; carrier risks indicated for each individual; f = general population carrier frequency

the proband and his spouse are both carriers is the carrier frequency in the population multiplied by two-thirds. The risk that their present pregnancy will produce an affected child is one quarter of this or 2/3 x f x 1/4 (where f is the general population carrier frequency).

X-linked recessive traits

Fig. 13.6 indicates an X-linked recessive pedigree. As indicated certain of the females are obligate carriers. For each obligate carrier on average one half of her boys will be affected and one half of her daughters carriers. If affected males reproduce then their sons will be normal but their daughters will be obligate carriers. The difficulty arises in the counselling of the non-obligate carrier females. Tests may be available for carrier detection but few of these are absolute and this information needs to be combined with the pedigree risk using Bayes' theorem (Appendix 2). Extended family studies may be required to counsel all females at risk.

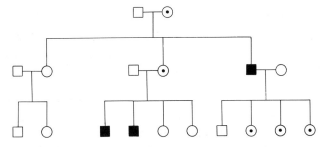

Fig. 13.6 X-linked recessive trait in a family; obligate carriers indicated

Multifactorial disorders

For discontinuous multifactorial traits empiric recurrence risks are used. These are simply the observed (rather than the calculated) recurrence risk for different relatives of an affected individual. Strictly empiric recurrence risks apply only to the population from which they were collected.

Infertility

One in ten of all couples are involuntarily infertile. As part of the investigation of such a couple chromosomal analysis of both partners is indicated to exclude a balanced structural rearrangement and Klinefelter's syndrome.

Recurrent miscarriages

One in six of all recognised pregnancies ends as a spontaneous miscarriage. If a couple have had three or more first trimester spontaneous miscarriages then chromosomal analysis is indicated as in 3-5% of cases one partner will have a balanced structural rearrangement.

Consanguinity

A consanguineous couple are at increased risk for autosomal recessive conditions. If the family history is otherwise negative the risk of an abnormal pregnancy is doubled from the general population risk to one in 20. No special screening is required prior or during a pregancy unless indicated by other factors such as ethnic origin.

For a consanguineous mating with a known autosomal recessive condition in a relative the recurrence risk can be calculated from the proportion of genes in common (Appendix 3).

Exposure to irradiation or mutagens

In the male diagnostic X-rays are of little genetic significance. Long term there is an increase in autosomal dominant point mutations after gonadal irradiation but the overall risk is probably only 1 in 500.

The female oocyte is especially radiosensitive around the time of fertilization. Outside this period the risks are similar to or less than for the male. As a safety measure non-urgent X- rays in the female are performed during the first 10 days after day one of the last menstrual period. An accidental diagnostic X-ray (of one rad or less) during the early stages of pregnancy results in a total added risk of 1 in 1000 to the fetus for congenital malformation, mental retardation or cancer in childhood. Neither termination of pregnancy nor amniocentesis are indicated. The fetal risks increase in relation to the dose of X-rays and termination is generally advised if a fetus less than 8 weeks is exposed to more than 25 rads.

Chapter 14
Chromosomal Disorders

Chromosomal disorders include all conditions associated with visible changes of the chromosomes. Approximately 7.5% of all conceptions have a chromosomal disorder but most of these are spontaneously aborted so the birth frequency is 0.6% (Fig. 14.1). Amongst early spontaneous abortions the frequency of chromosomal disorders is 60% whereas in late spontaneous abortions and stillbirths the frequency is 5%. The types of chromosomal abnormality also differ within these different groups. Table 14.1 shows the types of chromosomal abnormality seen in early spontaneous abortions. With the exception of chromosomes 1 and 5 each type of autosomal trisomy has been seen and trisomy 16 is especially frequent. In contrast trisomy 16 is never seen in the

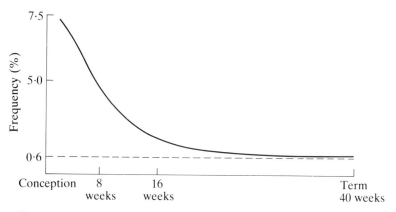

Fig. 14.1 Frequency of chromosomal abnormalities

Table 14.1 Chromosomal findings in early spontaneous abortions

40% apparently normal		
60% abnormal —	trisomy	30%
	45,X	10%
	triploid	10%
	tetraploid	5%
	other	5%

Table 14.2 Chromosomal disorders in newborns

Disorder	Birth frequency
Balanced translocation	1/500
Unbalanced translocation	1/2000
Pericentric inversion	1/100
Trisomy 21	1/700
Trisomy 18	1/3000
Trisomy 13	1/5000
47,XXY	1/1000 males
47,XYY	1/1000 males
47,XXX	1/1000 females
45,X	1/2500 females

newborn. Triploid fetuses may survive to term but the majority abort. Thus in general the chromosomal abnormalities which cause early spontaneous abortion tend to be those with the most severe effects on the fetus. Sex chromosome abnormalites are rare amongst early abortuses with the notable exception of 45,X.

Table 14.2 indicates the commonest chromosomal disorders seen in newborns. Not all of these chromosomal changes are associated with disease but in general autosomal abnormalities tend to be more severe than sex chromosomal abnormalities and deletions tend to be more severe than duplications. In the autosomal abnormalities developmental retardation, multiple congenital malformations and dysmorphic features are usual. Although the exact pattern of features may suggest the chromosomal disorder, no individual clinical feature is pathognomonic for a given chromosomal abnormality.

TRANSLOCATIONS

A translocation is the transfer of chromosomal material between chromosomes. Three main types are identifed: reciprocal, centric fusion and insertional (Chapter 5). Centric fusion translocations only involve the acrocentric chromosomes 13-15, 21 and 22 whereas reciprocal translocations may involve any of the chromosomes including the sex chromosomes. If the transfer results in no overall loss or gain of chromosomal material then the person will be healthy and is said to have a balanced translocation. The birth frequency for balanced translocations is one in 500 with approximately equal numbers of the centric fusion and reciprocal types but relatively few insertional translocations.

The carrier of a balanced translocation can be reassured that health and lifespan will be unaffected. Problems may arise, however, during meiosis with production of chromosomally unbalanced offspring. Some of these fetuses will spontaneously

Table 14.3 Risks of chromosomally unbalanced offspring for carriers of balanced structural rearrangements

Rearrangement	Carrier	Risk of unbalanced offspring
Centric fusion 13;14	Father	1%
Centric fusion 13;14	Mother	1%
Centric fusion 14;21	Father	1%
Centric fusion 14;21	Mother	10%
Centric fusion 21;22	Father	5%
Centric fusion 21;22	Mother	10%
Centric fusion 21;21	Father	100%
Centric fusion 21;21	Mother	100%
Reciprocal (any)	Father	10%
Reciprocal (any)	Mother	10%

abort but if liveborn they show multiple dysmorphic features and mental retardation. On theoretical grounds the majority of offspring should be unbalanced but the actual risk is much lower because of embryonic inviability and gametic selection (Chapter 5). The risk for unbalanced offspring depends upon the type of translocation and which parent is the carrier (Table 14.3). Reassurance may be given in a pregnancy at risk by fetal karyotyping after amniocentesis or chorion biopsy. A family study should also be undertaken to determine other outwardly healthy carriers who will be at similar reproductive risk. A de novo translocation where both parents have normal chromosomes is usually not associated with clinical abnormality. Occasionally, however, genes may be damaged at the breakpoints and produce chromosomal imbalance and an abnormal phenotype. This is a cause for concern when a de novo translocation is an incidental finding at amniocentesis.

PERICENTRIC INVERSIONS

Paracentric (ie. excluding the centromere) inversions do not produce a clinical abnormality in the carrier nor do they constitute an indication for prenatal diagnosis because if a crossover occurs within the inversion loop no viable gamete is produced (Chapter 5).

Pericentric inversion carriers are not clinically abnormal but there is a risk of producing chromosomally unbalanced offspring particularly when the inversion involves a large part of the chromosome. This risk is 1 in 10 for a carrier mother and 1 in 20 for a carrier father. Small pericentric inversions of chromosome 9, which are found in 1% of the population, are an exception to this and no chromosomally abnormal offspring resulting from crossover within the inversion have ever been described.

TRISOMY 21 (Mongolism, Down's syndrome)

Incidence
The overall birth incidence is 1 per 700 livebirths. The incidence at conception is greater but more than 60% are spontaneously aborted and at least 20% are stillborn. The incidence increases with increasing maternal age (Fig. 14.2). Thus the incidence at the 16th week of pregnancy (usual time for amniocentesis) is 1 in 200 for a 36 year-old mother rising to 1 in 100 at 39 years and 1 in 50 at 42 years.

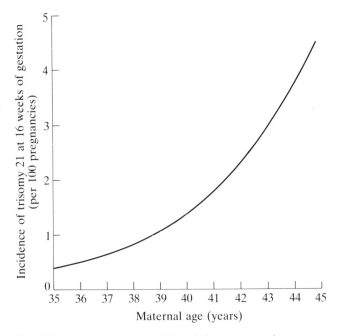

Fig. 14.2 Frequency of trisomy 21 in relation to maternal age

Clinical Features
The facial appearance often permits a clinical diagnosis. The palpebral fissures are upslanting with speckling of the iris (Brushfield's spots), the nose is small and the facial profile flat (Fig. 14.3). In the neonate hypotonia may be marked and redundant folds of skin about the neck are a feature of this and several other chromosomal disorders. The skull is brachycephalic with misshapen, low-set ears. A single palmar crease (Simian crease) may be present (50%) and the little fingers are short and incurved (clinodactyly, 50%). A wide gap between the first and second toes may be present.

Mental retardation is the most serious complication. The IQ is usually less than 50 and if not mosaicism should be suspected. Congenital heart malformations, especially endocardial cushion defects, are present in 40% and duodenal atresia may occur. Other

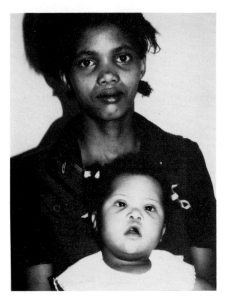

Fig. 14.3 Trisomy 21 phenotype

complications include cataracts (2%), epilepsy (10%), hypothyroidism and leukaemia (1%).

When serious malformations are present death during infancy is usual but otherwise life expectancy is little reduced. Trisomy 21 accounts for about one third of all moderate and severe mental handicap in children of school age. Most will walk and develop simple language. Puberty is often delayed and incomplete with adult heights about 150 cm.

Aetiology
Most cases (95%) are regular trisomy 21 (Fig. 14.4). This arises from non-disjunction at the first or second meiotic divisions. Overall the mother contributes the extra chromosome in 80% of cases and the father in 20%. Klinefelters syndrome coexists in 0.25% as a result of double non-disjunction. One percent of patients have mosaicism with normal and trisomy 21 cell lines. This arises after fertilization and the clinical features tend to be milder than in the full syndrome. In 4% of cases the child received the extra copy of chromosome 21 from a parent who is a carrier of a balanced translocation involving chromosome 21 (see previous section and Fig. 5.6) or has a de novo translocation.

Recurrence risk
For young parents who have produced a child with trisomy 21 the risk of recurrence of trisomy 21 or other major chromosomal abnormality is 1%. This is a low risk but many couples still seek reassurance by fetal karyotyping in future pregnancies. When the

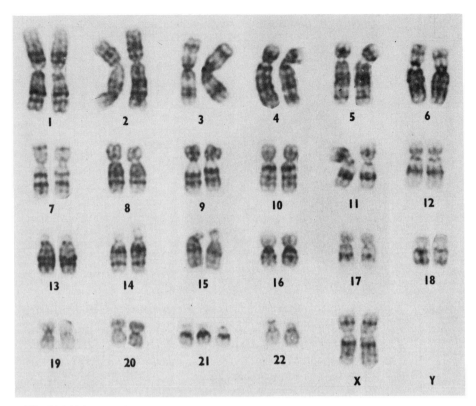

Fig. 14.4 Trisomy 21 karyotype

mother is over 35 years of age the age-specific risk should be used
(Fig. 14.2). The parents of a child with mosaic trisomy 21 have only
the age-specific risk of recurrence. The recurrence risks for carriers
of a balanced translocation are given in Table 14.3.

Affected persons rarely reproduce. The males are generally
infertile. For females with trisomy 21 one half of their offspring
would be expected to be affected.

47,XYY

Incidence

The incidence of 47,XYY is 1 per 1000 male births with no apparent
parental age effect. The frequency is increased in males in penal
institutions for the mentally subnormal (20/1000) and in mentally
deficient adult males (3/1000).

Clinical features

This chromosome disorder is often asymptomatic although
intelligence tends to be 10-15 points less than the normal siblings
and behaviour problems with aggression may occur. Patients tend

to be tall but have normal body proportions and no other clinical signs.

Aetiology
47,XYY arises form the production of a YY sperm at the second paternal meiotic division.

Recurrence risk
The recurrence risk is probably not increased for the parents of an affected child. For a person with 47,XYY the expected offspring would be 2XXY : 2XY : 1XX : 1XYY. In practice, fertility appears unimpaired in most cases yet only XX and XY offspring have been observed.

47,XXY (Klinefelter's Syndrome)

Incidence
Overall the birth incidence of 47,XXY is 1 per 1000 males with an increased risk at increased maternal age. The frequency is increased in infertile males (100/1000) and in males in institutions for the mentally retarded (10/1000).

Clinical Features
The diagnosis is generally made during adult life at the investigation of infertility since this is the single most common cause of hypogonadism and infertility in men. The testes are small (less than 2cm long in the adult) and fail to produce adult levels of testosterone. This leads to poorly developed secondary sexual characteristics and gynaecomastia (40%). The limbs are elongated from early childhood and the upper to lower segment ratio is abnormally low (Fig. 14.5). Scoliosis, emphysema, diabetes mellitus (8%) and osteoporosis may occur.

Testosterone replacement therapy will improve secondary sexual characteristics but infertility is the rule except in mosaics. Intelligence is 10-15 IQ points below the normal sibs with 20% mildly mentally retarded.

Aetiology
The extra X chromosome is of maternal origin in 60% and paternal in 40%. It may arise by non-disjunction at either the first or second maternal meiotic divisions but in the male can only arise when the first meiotic division produces an XY sperm. Studies of the Xg blood group may help to identify the parental origin of the non-disjunction. About 15% are mosaics 46,XY/47,XXY.

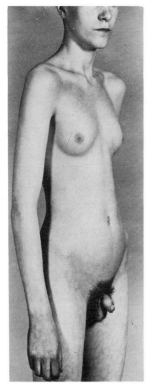

Fig. 14.5 Klinefelter's syndrome

Recurrence Risk
The recurrence risk is not increased above the general population risk after an affected child.

47,XXX

Incidence
The birth frequency is one per 1000 females with a maternal age effect.

Clinical Features
Individuals appear clinically normal but 15-25% are mildly mentally retarded.

Aetiology
Non-disjunction at either female meiotic division or at the male second meiotic division is the cause of 47,XXX.

Recurrence risk

The recurrence risk does not appear to be increased above the general population incidence. About three-quarters of affected females are fertile. One half of their offspring would be expected to be affected but in practice they are usually normal.

45,X (Turner's syndrome)

Incidence
The overall incidence is 1 per 2500 female births. The frequency at conception is much higher but 99% spontaneously abort.

Clinical Features
The diagnosis may be suggested in the newborn by redundant neck skin and peripheral lymphoedema (Figs. 14.6 and 14.7). Often the diagnosis is only made later during the investigation of short stature or primary amenorrhea.

Proportionate short stature is apparent from early childhood, no adolescent growth spurt occurs and adult heights lie between 50-60 inches (125-150cm). The chest tends to be broad (60%) with the impression of widely spaced nipples. The hair line is low (73%) and the neck may be webbed (54%) (Fig. 14.8). The carrying angle may be increased (56%) and the fourth metacarpals short (44%). Hypoplasia of the nails and multiple pigmented naevi are common. Peripheral lymphoedema occurs at some stage in 40%. The ovaries are represented by streaks and this results in failure of secondary sexual development.

Congenital heart disease, notably coarctation of the aorta and atrial septal defect, is present in 20% and there is also an increased risk of unexplained systemic hypertension (27%) and Hashimoto's thyroiditis. Intelligence and lifespan are normal. Sex hormone replacement will allow the development of secondary sexual characteristics but does not influence the ultimate stature nor the infertility.

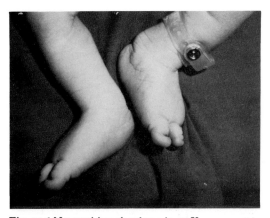

Fig. 14.6 Neonatal lymphoedema in 45,X

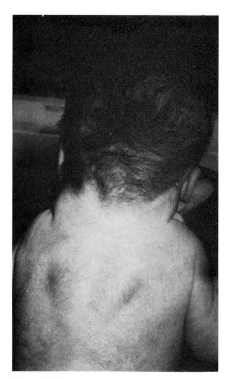

Fig. 14.7 Redundant neck skin in 45,X

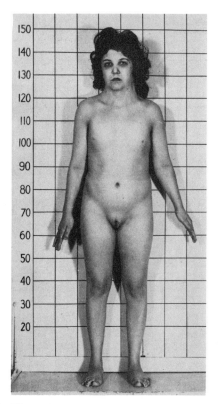

Fig. 14.8 Adult female with Turner's syndrome

Aetiology

Monosomy X may arise from non-disjunction in either parent. In 75% with monosomy X only the maternal X chromosome is present and thus the error occured in spermatogenesis. Overall 57% of patients have 45,X, 17% have an isochromosome of the long arm of X, 16% are mosaics, and 10% have a short arm deletion of one X. In general, deletion of the short arm of the X is associated with the Turner phenotype, while long arm deletions alone produce streak ovaries without the associated dysmorphic features.

Recurrence Risk

The recurrence risk does not appear to be increased above the negligible general population risk.

TRISOMY 18 (Edward's syndrome)

Incidence

The incidence of trisomy 18 is 1 per 3000 livebirths with a maternal age effect. The incidence at conception is much higher but 95% of affected fetuses spontaneously abort. At birth there is a preponderance of females which may reflect an excess of spontaneous abortion in the affected males.

Clinical Features

The birth weight is low and multiple dysmorphic features are apparent in the newborn. These include: a characteristic skull shape with a small chin and prominent occiput, low set malformed ears, clenched hands with overlapping index and fifth fingers, single palmar creases (30%), rockerbottom feet and a short sternum (Fig.14.9). The dermatoglyphics show a predominance of arches and in males cryptorchidism is usual.

Malformations of the heart, kidneys and other organs are frequent and 30% die within a month. Only 10% survive beyond the first year and these infants show profound developmental delay.

Aetiology

Parental non-disjuntion at either the first or second meiotic divisions results in the extra copy of chromosome 18. Rarely a parental translocation is responsible. Occasionally mosaicism is seen with a milder phenotype.

Recurrence risk

For parents of a child with regular trisomy 18 the recurrence risk is less than 1%.

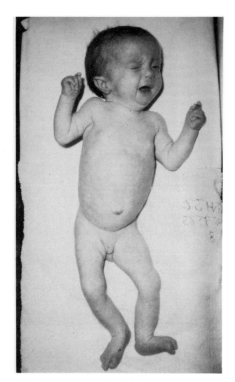

Fig. 14.9 Trisomy 18 phenotype

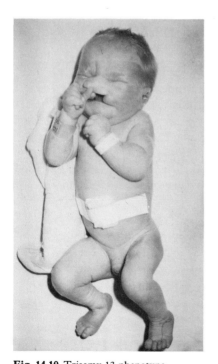

Fig. 14.10 Trisomy 13 phenotype

TRISOMY 13 (Patau's syndrome)

Incidence

The incidence of trisomy 13 is one per 5000 livebirths with a maternal age effect.

Clinical Features

Multiple dysmorphic features are apparent at birth. These include: hypotelorism reflecting underlying holoprosencephaly, microphthalmia, cleft lip and palate, abnormal ears, scalp defects, redundant skin about the nape, clenched fists, single palmar creases (60%), postaxial polydactyly, prominent heels and cryptorchidism in the male (Fig. 14.10).

Congenital heart disease is usual and 50% die within a month. Only 10 % survive beyond the first year.

Aetiology

Non-disjunction at either the first or second meiotic divisions in either parent may cause trisomy 13. In about 20% of cases one parent is a translocation carrier. In about 5% mosaicism is present.

Recurrence Risk

The recurrence risk is less than 1% provided a parent is not a carrier of a balanced translocation.

TRIPLOIDY

Incidence

Triploidy occurs in 2% of all conceptions but early spontaneous abortion is usual and survival to term exceptional.

Clinical Features

The neonate with triploidy has multiple abnormalities: marked low birth weight, disproportionately small trunk to head size, syndactyly, multiple congenital abnormalities and a large placenta with hydatidiform-like changes in most cases.

Aetiology

In most cases the extra set is paternally derived with 66% due to double fertilization, 24% due to fertilization with a diploid sperm and 10% due to fertilization of a diploid egg. Sixty percent are

69,XXY and most of the remainder are 69,XXX. Hydatidiform change is found only when there is a double paternal contribution.

Recurrence Risk
The recurrence risk is not known but is probably not increased.

FRAGILE X ASSOCIATED MENTAL RETARDATION (Martin Bell syndrome)

Clinical features
The key features are mental retardation, enlarged testes and an X chromosome fragile site. The testicular volume may be increased before puberty but is 30-50 ml in 50% of affected adults (normal 20 ml). Mental retardation is mild to moderate with a tendency for repetitive speech. Mental retardation of some degree is present in 20-30% of female heterozygotes.

In an affected male 10-60% of cells show a fragile site on the X chromosomes at Xq27 (Fig. 14.11). The carrier females, especially those with mental retardation, may also show a small percentage of cells with this fragile site but this is more difficult to demonstrate in older carriers (over 35 years).

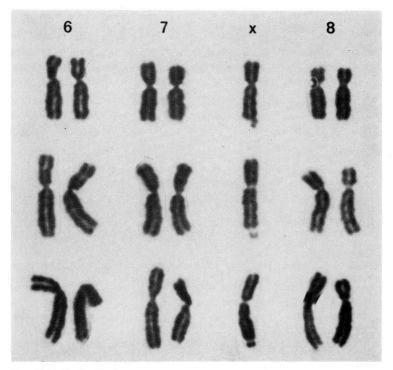

Fig. 14.11 Xq fragile site

Genetic aspects

X-linked mental retardation affects 1 in every 1000 boys and in one half of these families it is associated with the fragile X. Fragile X associated mental retardation is thus the second commonest identifiable cause of moderate and severe mental retardation in males after trisomy 21. Manifesting heterozygotes account for 7% of mild and 1% of moderate and severe mental retardation in females. Counselling is as for an X-linked recessive trait and prenatal diagnosis requires fetal sexing followed by fetal blood sampling in male pregnancies to demonstrate the fragile X. Exclusion of the carrier state, especially in older females, is not possible with current cytogenetic tests.

DELETIONS AND DUPLICATIONS

A large number of chromosomally unbalanced children have been described with visible deletions, duplications or combinations of the two. Overall the birth incidence is 1 per 2000.

These may arise from meiosis where a parent has a balanced structural rearrangement or as a new mutation. Any visible chromosomal imbalance of the autosomes almost invariably produces an abnormal phenotype with multiple dysmorphic features and mental retardation. The clinical features tend to be fairly non- specific although similar in siblings with the same aberration. The presence of duplicated or deleted genes in the involved areas of the chromosomes may be confirmed by gene dosage studies. These not only prove the chromosomal imbalance but also aid gene localisation (Chapter 8). For any of these chromosomal duplication-deletions, parental chromosomes must be performed to exclude balanced structural rearrangements. If parental chromosomes are normal then the recurrence risk is not increased above the general population incidence. If a parent is a balanced carrier then recurrence risks are given in Table 14.3.

PRADER-WILLI SYNDROME

Clinical features

In the newborn hypotonia and poor swallowing may be marked. The face is flat with a tented upper lip and the external genitalia are hypoplastic. In later childhood the hypotonia improves and overeating with obesity occurs. The forehead tends to be prominent with bitemporal narrowing. The palpebral fissures are almond-shaped and the hands and feet are small (Fig. 14.12). Mental retardation is universal.

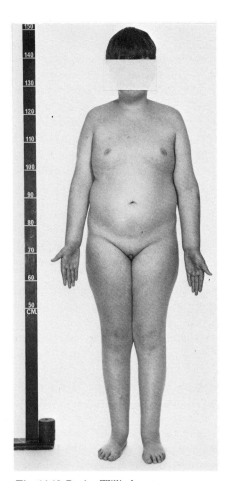

Fig. 14.12 Prader-Willi phenotype

Genetic aspects
In about 50% of patients with the Prader-Willi syndrome a small deletion is apparent in the long arm of chromosome 15 (at 15q11-13). In the remainder the chromosomes show no visible change. The empiric recurrence risk is 1.6% and prenatal diagnosis is not possible.

OTHER MICRODELETION DISORDERS

In addition to Prader-Willi syndrome five other conditions have so far been shown to be due on occasion to a visible microdeletion (Table 14.4).

The smallest visible deletion involves 6000 kilobases and thus large numbers of genes can be lost or gained without producing visible changes. The clinical effects of such abnormalities would be mental retardation, multiple congenital abnormalities and/or

Table 14.4 Conditions which may be caused by visible chromosomal microdeletions

Condition	Site of microdeletion
Prader-Willi syndrome	15q11
DiGeorge syndrome	22q11
Aniridia	11p13
Multiple endocrine neoplasia II	20p12
Retinoblastoma	13q14
Langer-Giedion syndrome	8q23

dysmorphic features. In children with these clinical features but normal chromosomes on light microscopy other methods need to be developed to demonstrate chromosomal imbalance. Gene dosage as described above would confirm the change but the problem lies in knowing which of several thousand gene products to measure.

Chapter 15
Single Gene Disorders

To date more than 3500 human single gene disorders have been described. Individually these are rare but collectively they affect about 1% of the population. Their importance in genetic counselling is threefold. First, it is vital to provide the affected families with correct advice despite the rarity of the condition. Second, these high risk rare conditions may be confused with commoner low risk causes of diseases such as cancer, ischaemic heart disease, diabetes mellitus and epilepsy. These and other heterogeneous (ie. multiple different causes) conditions of adult life are covered from the medical genetic aspect in Chapter 16. Finally, different single gene disorders can produce a similar phenotype (genetic heterogeneity). Genetically heterogeneous conditions may thus have different modes of inheritance in different families and this poses problems when counselling. Sometimes in a family it is impossible to determine which of two or more modes of inheritance may be operative and then a combined (composite) risk figure is given.

Conditions which are notorious for their genetic heterogeneity are considered first and then the commoner and clinically most relevant single gene disorders are outlined. These outlines should serve to allow the reader to select appropriate reference texts for further information.

CONDITIONS WITH MARKED GENETIC HETEROGENEITY

Mental retardation

Intelligence is determined as a multifactorial trait (Chapter 9). Some 3.4% of the population are mentally retarded with an IQ of less than 70. The majority of these individuals (3%) have an IQ between 50-70 (mild mental retardation) and these individuals show twin concordance and family correlations which support multifactorial inheritance. Occasionally a single gene (eg. neurofibromatosis) or chromosomal (eg. 47,XXY) disorder is responsible.

Moderate and severe mental retardation (IQ < 50) affects 1% of

Table 15.1 Causes of moderate and severe mental retardation

40% chromosomal — trisomy 21 (32%)
                   — other autosomal (2%)
                   — sex chromosomal abnormality (6%)
15% monogenic
2% maternal infection
7% perinatal asphyxia
2% postnatal infection or trauma
34% unknown — with congenital defect or dysmorphic features (14%)
                — with additional evidence of brain damage (10%)
                — without additional abnormality (10%)

newborns but this incidence falls to 0.3-0.4% in older children due to deaths in infancy. The cause can be defined in about two-thirds of cases (Table 15.1). The single commonest cause is trisomy 21 and this is followed in males by mental retardation associated with the fragile X (Chapter 14).

When chromosomes are normal and no other cause can be identified then the recurrence risk depends upon the sex of the proband. The risk to sibs is 1 in 10 for a male proband and 1 in 25 for a female proband. This imbalance probably reflects as yet undelineated forms of X-linked recessive mental retardation. If two sibs are affected then a recessive trait, either autosomal or X-linked is likely. Similarly, if a male proband has affected maternal male relatives X-linked inheritance must be considered.

Congenital deafness

One in every 1000 children is congenitally deaf. The cause is genetic in 50% of cases but different modes of inheritance are apparent in different families (Table 15.2). The situation is even more complicated as several non-allelic autosomal recessive types are recognised. Overall 20% of the families of two congenitally deaf parents will contain deaf offspring; in half of these only some will be deaf (autosomal dominant inheritance) whereas in the others all children will be deaf (allelic autosomal recessive inheritance).

For normal parents of a congenitally deaf child with no identifiable environmental cause the risk of recurrence is 1 in 6. This composite figure is derived from the proportion of autosomal dominant families with a negligible new mutation risk of recurrence

Table 15.2 Causes of congenital deafness

50% monogenic — 87% autosomal recessive
12% autosomal dominant
1% X-linked recessive
30% environmental
20% idiopathic

added to the 1 in 4 risk for the proportion of families with autosomal recessive inheritance. If a second child is deaf then autosomal recessive inheritance is accepted and the recurrence risk is modified to 1 in 4.

Congenital cataract

One in every 250 neonates has congenital cataract. The cause is heterogeneous (Table 15.3). For the normal parents of an affected child the composite recurrence risk is 1 in 10. As for congenital deafness if recurrence does occur then autosomal recessive inheritance is accepted.

Table 15.3 Causes of congenital cataract

22% chromosomal
25% prenatal infection
9% metabolic
10% genetic — most autosomal dominant
— some autosomal recessive
34% idiopathic

Retinitis pigmentosa

Retinitis pigmentosa results in progressive visual loss associated with clumps of retinal pigment resembling bone corpuscles. Early there is loss of night vision with constricted visual fields but later central vision may also be impaired. Atypical forms may accompany diseases such as myotonic dystrophy and Freidreich's ataxia.

The typical form may be autosomal dominant (15%), autosomal recessive (80%) or X-linked recessive (5%) with an overall prevalence of 1/5000. Counselling may be easy if other family members are affected but as the different types are clinically indistinguishable composite risks are necessary for isoated cases. The composite risk for the child of an isolated case is 1 in 8 but if the child is affected then autosomal dominant inheritance is accepted.

The X-linked form is closely linked to the DNA probe L1.28 (Chapter 7) and this should permit both carrier detection and prenatal diagnosis. Prenatal diagnosis for the other types is not possible.

Osteogenesis imperfecta

Osteogenesis imperfecta is characterised by a combination of the following: brittle bones, blue sclerae, discoloured teeth, and deafness due to otosclerosis. Overall incidence is 1/10,000. The clinical severity shows marked variation from no symptoms to perinatal death (Fig. 15.1). In intermediate cases multiple fractures and bowing lead to physical disability.

Genetic heterogeneity is marked. The lethal and intermediate

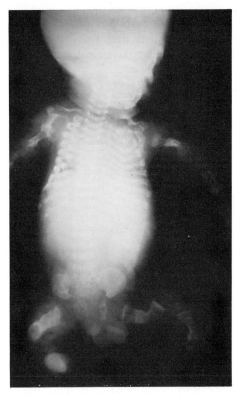

Fig. 15.1 Perinatal-lethal variant of osteogenesis imperfecta

forms may be either autosomal recessive or dominant. The mild
variant is usually autosomal dominant. Pathological heterogeneity
at the molecular level reflects the clinical variability (Chapter 12).
Prenatal diagnosis of the lethal variants is possible with ultrasound
scanning and may be possible for the variants with known
molecular pathology.

Arthrogryposis

Arthrogryposis or multiple non-progressive joint contractures
affects 1 in every 5000 births. Identifiable causes include trisomy 18,
neural tube defect, autosomal dominant distal arthrogryposis, and
idiopathic anterior horn cell loss. Despite the heterogeneity, for
normal parents the composite recurrence risk is small.

Disproportionate dwarfism

Disproportionate short stature may be either short-trunked or
short-limbed. Mucopolysaccharidosis type IV is the commonest
cause of short-trunked and achondroplasia the commonest cause of
short-limbed disproportionate short stature but numerous other
genetic and non-genetic conditions can produce such a change in
body proportions (see specialised texts- reference section).

Achondroplasia

Diagnosis
Short limbs especially proximally; normal length trunk with lumbar scoliosis; prominent forehead with depressed nasal bridge; trident hand; X-ray caudal narrowing of lumbar interpedicular distance (Fig. 15.2).

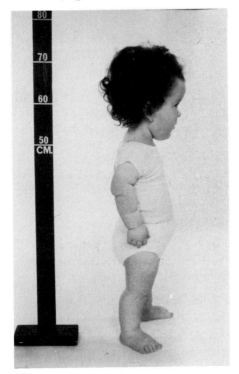

Fig. 15.2 Achondroplasia

Prognosis
Adult height 132 cm in male, 123 cm in female; normal IQ and lifespan; backache common; spinal cord compression rare.

Genetics
Autosomal dominant trait with full penetrance and little variation in expressivity; incidence 1/26,000 livebirths; mutation rate 14 per million gametes with a paternal age effect.

Alpha-1-antitrypsin deficiency

Diagnosis
Assay of protease inhibitor activity and typing by isoelectric focussing.

Prognosis
Only ZZ homozygotes affected; juvenile cirrhosis (10%); pulmonary emphysema in 3rd or 4th decade (60-70%).

Genetics
Autosomal codominant trait with 23 alleles at a locus on chromosome 14; 88% of population MM, 3.5% MZ and 0.06% ZZ.

Androgen insensitivity syndrome (Testicular feminisation syndrome)

Diagnosis
Female phenotype with normal breast development but primary amennorhea and paucity of pubic and other body hair; blind vaginal pouch, intra-abdominal testes; 46,XY karyotype.

Prognosis
Inguinal herniae (50%); gonadal neoplasms if not removed (5-20%); infertility (100%); normal IQ and lifespan.

Genetics
X-linked recessive trait leading to a deficiency of cell binding of testosterone and dihydrotestosterone; carrier detection not practical; incidence 1/62,400 liveborn males.

Apert syndrome

Diagnosis
Craniosynostosis of several sutures; bony syndactyly of digits 2-5 (Fig. 15.3).

Prognosis
Mental retardation (?%); cleft palate (25%).

Genetics
Autosomal dominant; low biological fitness so most cases are new mutations; paternal age effect for new mutations.

Bloom syndrome

Diagnosis
Low birth weight; short stature, photosensitive facial rash, tenfold increase in the rate of sister chromatid exchanges.

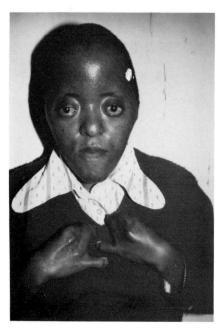

Fig. 15.3 Apert syndrome

Prognosis
Malignancy in the majority especially of the intestinal tract and leukaemia; mean age at death 19 years.

Genetics
Autosomal recessive with a maximal incidence in Ashkenazi Jews of 1 in 100,000; carrier detection not possible; prenatal diagnosis possible.

Congenital adrenal hyperplasia (21-hydroxylase deficiency)

Diagnosis
Neonatal vomiting, shock and death in the salt-losing form; virilisation of the female with ambiguous genitalia; precocity in the male; elevated urinary ketosteroids and pregnanetriol; markedly elevated serum 17 alpha- hydroxyprogesterone and ACTH; biochemical abnormalities restored to normal with therapy.

Prognosis
Normal lifespan, health and fertility if promptly diagnosed and replacement therapy with hydrocortisone and fludrocortisone instituted.

Genetics
Autosomal recessive trait with locus linked to HLA on chromosome 6; non-salt losing and salt-losing (30%) forms appear to be allelic;

carrier detection not possible except by HLA linhage studies but prenatal diagnosis availale by assay of 17 alpha-hydroxyprogesterone in amniotic fluid. Incidence 1/40,000 births in USA, 1/500 births in Yupik Eskimoes. Deficiency of 21-hydroxylase is the commonest cause (90%) but rarer enzyme deficiencies are known which have different clinical features and biochemical abnormalities.

Congenital spherocytosis

Diagnosis
Chronic haemolytic anaemia with splenomegaly and unconjugated hyperbilirubinaemia; spherocytes with increased MCHC; reduced red cell survival and increased osmotic fragility.

Prognosis
Splenectomy abolishes the need for repeated blood transfusion; normal lifespan.

Genetics
Autosomal dominant; incidence 1/4500 births.

Cystic fibrosis

Diagnosis
Sweat sodium and chloride exceed 60mEq/L; absent trypsin in pancreatic juice; elevated serum immunoreactive trypsin levels in newborn permits neonatal screening.

Prognosis
Pancreatic insufficiency; chronic lung disease secondary to recurrent infection; rectal prolapse (5-10%); meconium ileus (5-10%); cirrhosis of the liver (1-5%); median survival 19 years but variable clinical severity (?multiple allelism).

Genetics
Autosomal recessive, incidence 1 in 2000 in Northern Europe with carrier frequency of 1 in 23, less common in Negroes and Orientals; carrier detection not possible; prenatal diagnosis by altered amniotic fluid microvillar enzyme levels under evaluation.

Cystinuria

Diagnosis
Increased urinary excretion and reduced intestinal absorption of cystine, lysine, arginine and ornithine.

Prognosis
Recurrent calculi; normal lifespan with therapy.

Genetics
Autosomal recessive; carrier detection possible; prenatal diagnosis
possible on amniotic fluid ; incidence 1/10,000.

Ehlers-Danlos syndrome

Diagnosis
Numerous variants are known with variable combinations of lax
joints, hyperelastic skin, vascular fragility and poor wound healing
(Fig. 15.4).

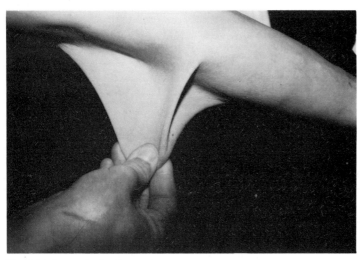

Fig. 15.4 Hyperelastic skin in Ehlers-Danlos syndrome

Prognosis
Lifespan is normal except in the variant with marked vascular
fragility.

Genetics
The commonest types are autosomal dominant but X-linked and
autosomal recessive forms are known; overall frequency 1/150,000
in the UK.

Friedreich's ataxia

Diagnosis
Progressive ataxia from 6-8 years; pes cavus; loss of deep tendon
reflexes in legs; extensor plantar response.

Prognosis
Progressive disability, chairbound in 2nd decade, death in 3rd decade.

Genetics
Autosomal recessive, carrier detection and prenatal diagnosis not possible. This is the commonest inherited cause of cerebellar ataxia but several other rarer genetic causes are known.

Galactosaemia

Diagnosis
Neonatal weight loss, vomiting, hepatomegaly, jaundice, cataracts, and susceptibility to infection; reducing substance in the urine (galactose); absent red cell galactose-1- phosphate uridyl transferase (GALT).

Prognosis
Exclusion of milk and milk products from the diet is necessary for life; with such a diet lifespan and IQ are normal however, mental retardation is invariable if therapy is delayed until after one month - hence the value of neonatal screening.

Genetics
Autosomal recessive trait; GALT locus on 9p; carrier detection possible (assay red cell GALT); prenatal diagnosis possible (assay of GALT in amniocytes); allelic Duarte variant with higher residual GALT activity has no symptoms even in the homozygote but may be symptomatic in the Duarte-galactosaemia genetic compound. Incidence of galacosaemia 1/40,000.

Gaucher disease

Diagnosis
Chronic adult type results in bone pain and splenomegaly; acute infantile form is a progressive neurological illness with hepatosplenomegly; intermediate variants known; reduced leucocyte beta glucosidase in all types.

Prognosis
1-2 years survival in infantile form; normal lifespan in adult form with supportive therapy.

Genetics
Autosomal recessive; heterozygote detection and prenatal diagnosis possible by beta glucosidase assay. Highest fequency in Ashkenazi Jews whose carrier frequency is 1 in 60.

Haemophilia A (Classical haemophilia)

Diagnosis
Recurrent haemorrhage postoperatively and spontaneously into soft tissues and joints; factor VIII less than 30% of normal (Fig. 15.5).

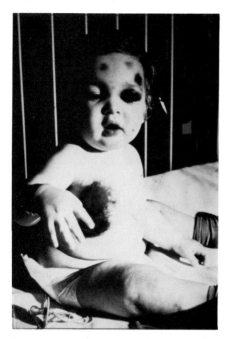

Fig. 15.5 Haemophilia A

Prognosis
With therapy lifespan near normal.

Genetics
X-linked recessive trait; 95% of obligate carriers can be distinguished from normal by comparing the ratio of factor VIII activity to cross-reacting factor VIII antigen level; prenatal diagnosis by fetal blood sampling and measurement of factor VIII level in male pregnancy; incidence 1/5000 male births. For an isolated case 90% of the mothers will be carriers.

Haemophilia B (Christmas disease)

Diagnosis
Clinically similar to haemophilia A; factor IX less than 30% of normal.

Prognosis
With therapy lifespan near normal.

Genetics
X-linked recessive trait; 25-30% of obligate carriers can be distinguished from normal by assay of factor IX; DNA diagnosis under evaluation for carrier detection; prenatal diagnosis possible by fetal blood sampling or DNA analysis; frequency 1/30,000 male births.

Hepatolenticular degeneration (Wilson disease)

Diagnosis
Defective copper metabolism with reduced serum caeruloplasmin and increased hepatic copper.

Prognosis
Untreated results in chronic active hepatitis and neurological sequelae; normal lifespan with therapy

Genetics
Autosomal recessive; carrier detection possible in 10%; prenatal diagnosis not possible; incidence 1 in 200,000.

Hereditary motor and sensory neuropathy type I (Peroneal muscular atrophy, Charcot-Marie-Tooth disease)

Diagnosis
Onset in teens with bilateral foot drop; enlarged peripheral nerves; inverted champagne bottle legs; pes cavus; reduced ankle jerks; some sensory loss, especially for vibration; reduced nerve conduction velocity; muscle biopsy shows demyelination.

Prognosis
Normal lifespan with mild to moderate disability.

Genetics
Usually autosomal dominant; rarely autosomal or X- linked recessive.

Holt-Oram syndrome (Hand-Heart syndrome)

Diagnosis
Variable arm malformations from hypoplastic thumb to phocome-

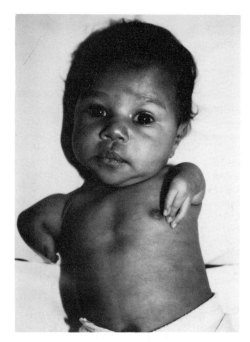

Fig. 15.6 Phocomelia in the Holt-Oram syndrome

lia; congenital heart disease especially atrial septal defect (Fig. 15.6).

Prognosis
Variable depending upon the severity of the malformations.

Genetics
Autosomal dominant with marked variation in expression; tends to be more severe in females.

Huntington chorea

Diagnosis
Onset usually in 4th decade; psychiatric symptoms; progressive chorea and dementia.

Prognosis
Progressive disability; death in 10-12 years from onset.

Genetics
Autosomal dominant with complete but age-dependent penetrance - 10% by 30 years, 30% by 40, 60% by 50, 85% by 60 and 95% by 70 years; incidence 1/18,000 in UK and USA, 1/333,000 in Japan, 1/5740 in Tasmania. Linkage to a chromosome 4 molecular marker under evaluaton for presymptomatic and prenatal diagnosis.

Hypercholesterolaemia (Hyperlipidemia II)

Diagnosis
Defective cell surface receptor for low density lipoprotein; onset in 3rd or 4th decade with xanthomata, xanthelasma, corneal arcus and increased serum cholesterol (Fig. 15.7).

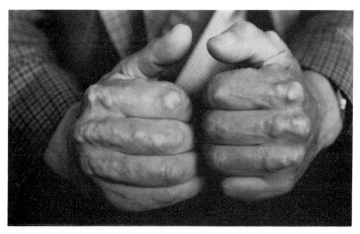

Fig. 15.7 Tendon xanthomata in familial hypercholesterolaemia

Prognosis
Premature death from ischaemic heart disease.

Genetics
Autosomal dominant; locus on chromosome 19; homozygous dominant has a disease of increased precocity and severity; prenatal diagnosis of homozygous dominant but not of heterozygote is possible.

Ichthyosis X-linked

Diagnosis
Commonest cause of ichthyosis in the first three months; in adults abdomen, legs and popliteal fossae are most often affected; markedly reduced level of steroid sulphatase in hair roots.

Prognosis
Normal intelligence and lifespan.

Genetics
X-linked recessive trait; steroid sulphatase locus near tip of the short arm of the X chromosome; female carrier detection may be possible with assay of steroid sulphatase in hair roots.

Incontinentia pigmenti

Diagnosis
Onset in infancy of vesicular skin rash followed by irregular whorled pigmentation; partial alopecia, hypodontia.

Prognosis
Mental retardation (30%); ocular problems (30%).

Genetics
X-linked dominant with in utero lethality for hemizygous male and thus a marked excess of affected females.

Intestinal polyposis

Diagnosis
Three types:
Type I (Polyposis coli) polyps confined to colon
Type II (Peutz-Jeghers syndrome) polyps throughout the intestines; melanin spots on lips and fingers.
Type III (Gardner syndrome) colorectal polyps; facial bone osteomas; epidermoid or sebaceous cysts.

Prognosis
Type I : death from malignancy unless colectomy
Type II: intestinal malignancy rare; intussusception; ovarian granulosa cell tumour (10-15%)
Type III: death from malignancy unless colectomy

Genetics
All autosomal dominant; no prenatal diagnosis possible; combined frequency 1/10,000.

Marfan syndrome

Diagnosis
Arachnodactyly; long limbs with reduced upper to lower segment ratio; lax joints; evidence of complications (Fig. 15.8).

Prognosis
Lens subluxation; aortic fusiform or dissecting aneurysm; average lifespan 40-50 years.

Genetics
Autosomal dominant with 15% of cases new mutations; prenatal diagnosis not possible.

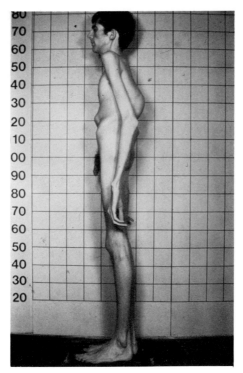

Fig. 15.8 Marfan syndrome

Mucopolysaccharidoses

Diagnosis

Four main types (although 7 types known):

Type I (Hurler syndrome) mucopolysacchariduria; deficiency of alpha-L-iduronidase

Type II (Hunter syndrome) mucopolysacchariduria; deficiency of sulpho-iduronide sulphatase.

Type III (Sanfillipo syndrome) mucopolysacchariduria; deficiency of heparan sulphate sulphatase or N-acetyl-alpha-D-glucosaminidase.

Type IV (Morquio disease) mucopolysacchariduria; deficiency of fibroblast 6-sulpho-N-acetylhexosaminido-sulphatatase

Prognosis

Type I coarse facies in infancy (100%); short stature (100%); progressive mental retardation (100%); cloudy corneae (100%); death in second decade (Fig. 15.9).

Type II as type I but later onset and clear corneae; death in third decade.

Type III progressive mental retardation in early childhood; normal facies, stature and corneae; death in second decade

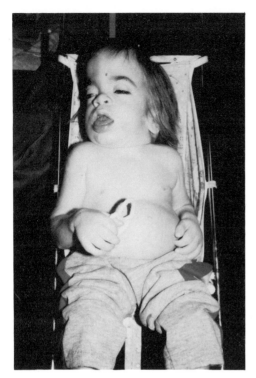

Fig. 15.9 Mucopolysaccharidosis type I (Hurler syndrome)

Type IV short stature with scoliosis; normal intelligence, facies, and corneae; atlantoaxial subluxation; death in third decade

Genetics
Combined frequency 1/20,000 with type III most common. Prenatal diagnosis possible by analysis of glycosaminoglycans in amniotic fluid by two-dimensional electrophoresis and by enzyme assay in cultured amniocytes; carrier detection not possible with the exception of Type II.

Multiple endocrine neoplasia

Diagnosis
Three types:
Type I - pituitary adenoma, hyperparathyroidism, pancreatic adenoma.
Type II - phaeochromocytoma, medullary thyroid carcinoma; hyperparathyroidism
Type III - neuromas of mucous membranes; blubbery lips; medullated corneal nerve fibres; megacolon; medullary thyroid carcinoma; phaeochromocytoma.

Prognosis
Variable for each depending upon malignancy or hyperfunction of the tumours.

Genetics
All autosomal dominant; prenatal diagnosis not possible; deletion of 20p12 found in some cases of MEN II.

Multiple exostoses (Diaphyseal aclasis)

Diagnosis
Multiple cartilage capped exostoses

Prognosis
Involved bones may be shortened; malignant change (2%)

Genetics
Autosomal dominant with 30% of cases new mutations; fully penetrant; incidence 1/2000.

Muscular dystrophy - X-linked

Diagnosis
Two types:
Duchenne muscular dystrophy - onset in early childhood of progressive proximal muscle weakness; calf pseudohypertrophy; marked elevation of serum creatine phosphokinase; abnormal electromyogram and muscle biopsy.
Becker muscular dystrophy - onset of progressive muscular weakness in late childhood; calf pseudohypertrophy; marked elevation of serum creatine phosphokinase; abnormal electromyogram and muscle biopsy.
Prognosis:
Duchenne - mild mental retardation (25%); chairbound about 10 years; death about 20 years
Becker - often chairbound about 25 years from onset; lifespan may be normal

Genetics
Both X-linked recessive; ? allelic disorders loci at Xp21; measurement of creatine phosphokinase will detect 67% of obligate carriers for Duchenne and 50% for Becker; prenatal diagnosis by selective termination of all at risk male pregnancies; incidence Duchenne 1/3000 male births, Becker 1/20,000. Duchenne muscular dystrophy is the commonest human X- linked lethal condition. Diagnosis by DNA analysis is currently under development.

Myotonic dystrophy

Diagnosis
Progressive muscle weakness in early adult life; myotonia.

Prognosis
Cataracts (85%); frontal baldness in males; gonadal atrophy; severe disability usual 15-20 years from onset.

Genetics
Autosomal dominant, locus on chromosome 19; prenatal dignosis possible in 12% of families by linkage to secretor locus. For an affected female - 50% of children unaffected, 29% affected in later life, 12% neonatal deaths, 9% severe neonatal hypotonia and mental retardation (Fig. 15.10); frequency 1/20,000.

Fig. 15.10 Mother and child with myotonic dystrophy

Nail-Patella syndrome

Diagnosis
Nail dysplasia; absent or hypoplastic patellae; iliac horns (pathognomonic).

Prognosis
Nephropathy (30-50%) may lead to renal failure otherwise normal lifespan.

Genetics
Autosomal dominant with full penetrance, locus on chromosome 9q34 linked to the ABO locus.

Neurofibromatosis (von Recklinghausen's disease)

Diagnosis
Multiple cafe-au-lait patches; skin neurofibromata; axillary freckling (Fig. 15.11).

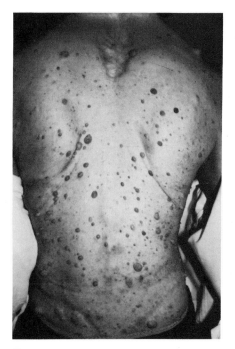

Fig. 15.11 Neurofibromatosis with scar from spinal cord decompression

Prognosis
Scoliosis; mild mental retardation (10%); seizures (3%); spinal cord or root compression; CNS tumour (5%); pseudarthrosis of the tibia (1%), phaeochromocytoma (<1%).

Genetics
Autosomal dominant with full penetrance; prenatal diagnosis not possible; frequency 1/3000.

Otosclerosis

Diagnosis
Progressive conductive hearing loss in middle age with normal tympanic membranes.

Prognosis
Hearing is restored by surgery

Genetics
Autosomal dominant with 25-40% penetrance; frequency 1/330 in Caucasians, 1/3300 in Negroes, 1/333,000 in Orientals.

Phenylketonuria

Diagnosis
Elevated blood and urine phenylalanine due to deficiency of hepatic phenylalanine hydroxylase.

Prognosis
Normal development and lifespan with low phenylalanine diet; mental retardation if untreated; risk of mentally retarded offspring for treated female unless diet is reintroduced prior to pregnancy.

Genetics
Autosomal recessive, locus on chromosome 12; carrier detection and prenatal diagnosis possible by DNA analysis; frequency 1/5000 livebirths in Scotland, 1/15,000 in USA, rare in Negroes and Ashkenazi Jews.

Retinoblastoma

Diagnosis
Usual onset in first two years with white cat's eye reflex or squint; bilateral in 20-30%

Prognosis
90% cured if unilateral and small

Genetics
Familial case are autosomal dominant with 90% penetrance, locus on chromosome 13q14; prenatal diagnosis not possible; incidence 1/18,000 births.

Spinal muscular atrophy

Diagnosis
Progressive weakness due to anterior horn cell loss; reduced or absent deep tendon reflexes; fasiculation; normal CPK; EMG shows denervation and muscle biopsy shows atrophy; infantile onset form (Werdnig-Hoffman) commonest with later onset variants.

Prognosis
Infantile form lethal at 1-2 years; juvenile forms may progress very slowly and produce little reduction in lifespan; IQ normal.

All variants autosomal recessive; collective frequency 1/25,000 births; prenatal diagnosis and carrier detection not possible. Carrier frequency 1 in 30 for infantile form, 1 in 90 for other types.

Tay-Sachs disease

Diagnosis
Progressive neurological abnormalities from late infancy; cherry-red macular spot; reduced serum hexosaminidase A.

Prognosis
Lethal by 3-4 years

Genetics
Autosomal recessive; carrier detection and prenatal diagnosis possible by assay of hexosaminidase A. Highest frequency in Ashkenazi Jews (1/3600 births), carrier frequency 1/30. Carrier frequency in non-Jews 1/300.

Treacher Collins syndrome (mandibulofacial dysostosis)

Diagnosis
Small mandible; malar hypoplasia; malformed ears (Fig. 15.12).

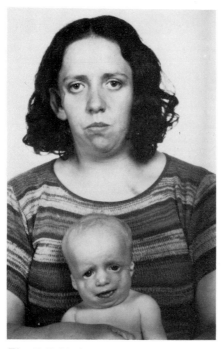

Fig. 15.12 Mother and child with Treacher Collins syndrome. Note variable expression

Prognosis
Conductive deafness (28%); cleft palate (32%); mental retardation (5%)

Genetics
Autosomal dominant with full penetrance but very variable expression; 60% of cases are new mutations.

Tuberous sclerosis

Diagnosis
White skin patches (82% by 5 years); facial fibroangiomatous rash (50% by 5 years); skin fibromatous plaques (shagreen patches); whitish retinal phakomata; intracranial calcification; evidence of complications (Fig. 15.13).

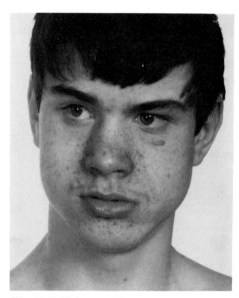

Fig. 15.13 Tuberous sclerosis

Prognosis
Epilepsy (90%); mental retardation (60%); variable lifespan which depends upon the severity of the retardation.

Genetics
Autosomal dominant with 85% of cases new mutations; prenatal diagnosis not possible

von Hippel-Lindau disease

Diagnosis
Retinal haemangioma; cerebellar haemangioblastoma.

Prognosis
Renal carcinoma (20%); renal, pancreatic, hepatic and epididymal cysts.

Genetics
Autosomal dominant; prenatal diagnosis not possible.

Waardenburg syndrome

Diagnosis
Lateral displacement of inner canthi; white forelock; heterochromia iridis; deafness.

Prognosis
Bilateral severe sensorineural deafness

Genetics
Autosomal dominant; prenatal diagnosis not possible; frequency 1/42,000.

Wilms tumour

Diagnosis
Solid renal mass in childhood with characteristic histopathology at nephrectomy; 10% bilateral.

Prognosis
Variable depending upon stage, worse if bilateral.

Genetics
One third are hereditary (especially if bilateral); autosomal dominant with 60% penetrance; incidence 1/10,000 births.

Xeroderma pigmentosum

Diagnosis
Failure to repair DNA damage after exposure to ultraviolet light.

Prognosis
Multifocal skin neoplasms; corneal scarring.

Genetics
Autosomal recessive; prenatal diagnosis possible; frequency 1/100,000; carrier detection not possible.

Chapter 16
Multifactorial Disorders

In contrast to single gene disorders, multifactorial conditions are determined by the summative effect of multiple genes at different loci together with environmental factors. As outlined in Chapter 9 multifactorial inheritance is operative for most normal characteristics in addition to many congenital malformations and common chronic diseases of adult life. Adult multifactorial disorders are discussed here whilst congenital malformations are covered in Chapter 17.

On the border between single gene and multifactorial traits are a group of single gene traits which are asymptomatic in the absence of an environmental trigger(s) (Table 16.1).

Table 16.1 Examples of single gene-environment interaction to produce disease

'At risk' genotype	Environmental trigger(s)	Disease
Glucose 6-phosphate dehydrogenase deficiency	certain drugs, broad beans, moth balls	Haemolytic anaemia
Uroporphyrinogen I synthetase deficiency	certain drugs	Porphyric crisis
Pseudocholinesterase deficiency	succinylcholine	Prolonged apnea
Autosomal dominant trait	halothane, succinylcholine	Malignant hyperpyrexia
Ornithine transcarbamylase deficiency	insect repellant	Coma

Glucose 6-phosphate dehydrogenase deficiency

Glucose 6-phosphate dehydrogenase activity is determined by a gene near the tip of the long arm of the X chromosome. This enzyme is important for the generation of NADPH an essential reducing agent for red cells. More than 150 molecular variants of this enzyme are known (multiple allelism). Some have normal enzyme activity whereas others result in reduced levels (Table

Table 16.2 Common glucose 6-phosphate dehydrogenase variants

Variant	Enzyme activity	Frequency
B	100%	normal
A	90%	1 in 5 US black males
A-	15%	1 in 10 US black males, 1 in 50 black females
B-	4%	common in Mediterranean area

Table 16.3 Environmental factors which can trigger a haemolytic episode in an individual with glucose 6-phosphate dehydrogenase deficiency

Drugs — antimalarials, sulphonamides, nitrofurantoin, aspirin, probenecid, chloramphenicol, quinidine
Chemicals — naphthalene
Other — broad beans (favism), infections

16.2). The highest frequencies of these variants are attained in areas where malaria is or was a problem. It is estimated that more than 100 million people in the World have a variant with reduced enzyme activity.

Despite the reduced level of activity in the hemizygous male there are usually no symptoms until the person is exposed to an environmental trigger (Table 16.3). Exposure to one of these agents results in an acute haemolytic anaemia. The heterozygous female is usually asymptomatic even with exposure to these agents. Family members at risk can be screened by quantitative and qualitative studies of red cell glucose 6-phosphate dehydrogenase.

Acute intermittent porphyria

This is an autosomal dominant trait which results in a reduction in the level of uroporphyrinogen I synthetase, an enzyme, which is produced by a gene near the tip of the long arm of chromosome 11. This is an exception to the rule that enzyme deficiencies are generally inherited as autosomal recessive traits.

Most persons are asymptomatic in the absence of an environmental trigger (Table 16.4). With exposure to one of these agents a porphyric crisis occurs with abdominal pain, vomiting, red urine, confusion, and neuropathy which can result in respiratory paralysis. Family members at risk should be screened with assay of red cell uroporphyrinogen I synthetase.

Table 16.4 Factors which can precipitate an acute attack of porphyria

Drugs — barbiturates, sulphonamides, griseofulvin, diphenylhydantoin, oestrogens
Other — infections, starvation

Succinylcholine sensitivity

Serum cholinesterase is an enzyme produced by a gene on chromosome 3. At this locus two codominant alleles occur Ei^u and Ei^a. Three genotypes are possible (Table 16. 5). Each of these genotypes is asymptomatic unless the individual is given a general anaesthetic. Succinylcholine (suxamethonium) is often used during the induction of general anaesthesia as a muscle relaxant. It is rapidly hydrolysed by serum cholinesterase except in the $Ei^a Ei^a$ individuals who have prolonged muscle paralysis and need artificial ventilation until recovery occurs. Family members at risk can be screened by quantitative and qualitative assay of serum cholinesterase.

Table 16.5 Serum cholinesterase genotypes

Genotype	Frequency
$E_1{}^u E_1{}^u$	96%
$E_1{}^u E_1{}^a$	4%
$E_1{}^a E_1{}^a$	1/20,000

Hyperthermia of anaesthesia (Malignant hyperpyrexia)

One in 20,000 of the population has this autosomal dominant trait. Individuals are asymptomatic until given a general anaesthetic. The trigger is either suxamethonium or halothane and this results in explosive hyperpyrexia, hypertonia, a massive elevation of creatine phosphokinase and death (60%). Family members at risk can be screened by muscle biopsy and in vitro exposure of the muscle to trigger agents.

COMMON CHRONIC DISEASES OF ADULTHOOD

Table 16.6 lists the major common chronic diseases of adulthood. Most if not all of these are heterogeneous and each can be shown to have a substantial multifactorial inherited component (Chapter 9). For example, epilepsy may be inherited as a single gene or multifactorial trait, may be a symptom of a chromosomal aberration or may be non-genetic in aetiology.

For the following multifactorial disorders the nature of the genetic component and of the environmental component is largely obscure. Some, like the above examples, may be the result of only a single locus with an environmental trigger whereas in others multiple loci each with a small but additive effect will be involved.

For each common and some of the less common adult disorders the relative importance of the genetic factors in aetiology will be indicated together with the empiric recurrence risks for genetic

Table 16.6 Common chronic diseases of adulthood

	Frequency /1000
Peptic ulcer	40-50
Rheumatoid arthritis	20
Diabetes mellitus — insulin dependent	2
— insulin independent	20
Ischaemic heart disease <65 years	10-20
Schizophrenia	10
Manic-depressive psychosis	4
Epilepsy	10
Others	4
TOTAL	@140/1000

counselling. Empiric recurrence risks are derived by observation of families with the conditions rather than from calculation as with single gene traits.

Cancer

Cancer is uncommon in childhood (1 in 160) but affects 1 in 10 of all adults at some stage in their lives. The aetiology is heterogeneous yet similar pathogenetic pathways seem to be involved (Chapter 12).

Although more than 150 single gene traits have cancer as a recognised complication these only account for a small proportion of all cancer. In each the cancer is of an unusual type or is associated with other phenotypic features (Table 16.7). Recognition of these rare forms of cancer is important for counselling and to identify other relatives at risk.

For most common cancers such as those of lung, prostate and bowel the frequency in monozygotic and dizygotic twins is equal and the risk to family members is not increased above the general population risk. Breast cancer is an exception especially if bilateral and of early onset. A woman whose mother and sister are affected has a lifetime risk of 30%; if her mother is unaffected and she has two affected sisters her risk is 15%. If undue clustering of cases of cancer occurs in a family chromosomal analysis to exclude a structural rearrangement in the region of an oncogene is indicated.

Table 16.7 Examples of single gene traits with cancer as a complication

Polyposis coli
Neurofibromatosis
Retinoblastoma
Wilms tumour
Multiple exostoses
von Hippel-Lindau disease
Multiple endocrine neoplasia
Multiple self-healing epithelioma
Xeroderma pigmentosum

Ischaemic heart disease

Coronary atherosclerosis is the commonest cause of death in adults. By 55 years of age 1 in 60 males and 1 in 90 females will have evidence of ischaemic heart disease. The prevalence rises rapidly thereafter. In families with young affected individuals familial hypercholesterolaemia should be excluded. The risks are otherwise only modestly increased with onset of ischaemic heart disease less than 55 years of age in either a parent or sibling (Table 16.8). For ischaemic heart disease of later onset the risk in relatives is probably not increased.

Table 16.8 Risk of death from ischaemic heart disease by 55 years of age in the relatives of probands with ischaemic heart disease of onset before 55 years of age.

	Male first degree relative	Female first degree relative
Male proband	1/12	1/36
Female proband	1/10	1/12

Systemic hypertension

Within the general population the distribution of systolic and diastolic blood pressure follow a normal or Gaussian curve. By convention young adults with values persistently above 140/90 are considered to be hypertensive. Blood pressure, especially the systolic, tends to increase with increasing age and so upper normal limits are age-dependent. With the standard upper normal limits at least 10% of the population are or will become hypertensive. Most individuals with sustained high blood pressure are asymptomatic and have no identifible cause (idiopathic hypertension). For this group multifactorial inheritance seems to be involved with an increased but ill- defined risk to sibs and offspring who therefore merit regular screening.

Diabetes mellitus

Diabetes mellitus is a heterogeneous condition. At least three clinical subtypes can be identified; insulin dependent (juvenile onset type); insulin independent (maturity onset type) and maturity onset diabetes of youth. The latter is inherited as an autosomal dominant trait whilst the risks to relatives for the former two types suggest multifactorial inheritance in at least a substantial proportion (Table 16.9). The figures in this Table relate to Western European populations. Certain ethnic groups have much higher or lower frequencies of diabetes. For example 30% of Pima Indians and 20% of Maltese are diabetic whereas less than 0.5% of

Table 16.9 Risks for the relatives of a diabetic proband

Type of diabetes	Overall frequency	Risk to sib	Risk to offspring
Insulin dependent	1/500	1/33	1/33
Insulin independent	1/50	1/10	1/10

Eskimoes are affected. These differences might be environmental (perhaps diet-related) or reflect differences in allele frequency.

Bronchial asthma

One in 25 of the population has bronchial asthma which is often accompanied by hay fever (1 in 5 of the population) or atopic eczema (1 in 25 of the population). Twin and family studies support multifactorial inheritance for extrinsic asthma with a risk to first degree relatives of 1 in 8. For relatives of a proband with intrinsic asthma the risk is 1 in 20.

Peptic ulcer

Peptic ulcer affects 1 in 25 men and 1 in 50 women at some stage of their lives. If there is unduly severe ulceration or associated endocrine abnormality, multiple endocrine neoplasia should be excluded. For other causes of peptic ulcer the risk to sibs and off-spring is 1 in 10. In some peptic ulcer families increased production of pepsinogen can be demonstrated in multiple individuals in a pattern consistent with autosomal dominant inheritance. Not all of these individuals have ulceration, however, and the environmental triggers in the affected group are not known but might include smoking and dietary habits.

Inflammatory bowel disease

Two subtypes of non-infectious inflammatory bowel disease are recognised: ulcerative colitis (frequency 1/1500) and Crohn's disease (frequency 1/5000). Multiple environmental agents are suspected (but not proven) to trigger these conditions. Twin and family studies suggest multifactorial inheritance with a risk to sibs and offspring of between 1 in 100 and 1 in 250.

Gluten enteropathy (Coeliac disease)

Malabsorption syndrome secondary to gluten intolerance affects 1/2000 individuals in Northern Europe. The recurrence risk in sibs and offspring is 1 in 33 for symptomatic disease although 1 in 10 have an abnormal jejunal biopsy. The risk for symptomatic disease is less than 1 in 100 for second degree relatives. Clearly gluten is the environmental trigger here and the association with certain HLA antigens might be a lead towards the genetic component.

Ankylosing spondylitis

This distinctive arthropathy may be isolated or complicate inflammatory bowel disease. Overall 2/1000 men and 0.2/1000 women are symptomatically affected. Whether isolated or secondary, 95% of those affected have HLA B27 as compared with 7% of the normal population. Not all with this allele develop the disease and environmental factors interacting with this allele or a closely linked gene (linkage disequilibrium) are suspected. For anyone with HLA B27 the risk or ankylosing spondylitis is 9% whereas for those without this allele the risk is less than 1%.

Rheumatoid arthritis

One in 50 of the population has rheumatoid arthritis with three females to each affected male. The disease is weakly associated with HLA DrW4 and DrW3. The risk to sibs and offspring is 1 in 20.

Psoriasis

One to 2% of the population have psoriasis. Studies of family correlation and twin concordance support multifactorial inheritance with a risk to sibs and offspring of one in 10.

Glaucoma

Primary open angle glaucoma affects 1/200 elderly people. The risk to sibs and offspring is 1 in 10 and these merit regular ophthalmological screening.

Multiple sclerosis

In the UK 1/2000 of the population are affected. Throughout the World the prevalence seems to increase with increasing latitude except in Japan where the disease is extremely rare. There is a weak association with HLA Dr2. The risk to sibs and offspring is 1/100.

Epilepsy

Epilepsy is extremely heterogeneous (Table 16.10). One in 200 of the population has grand mal epilepsy. Where no cause can be identified the recurrence risk for sibs and offspring is 1 in 25.

Petit mal is less common and is typically associated with a 3 cycle-

Table 16.10 Causes of epilepsy

Cerebral malformation
Trauma
Neoplasm
Single gene trait
Chromosomal aberration
Multifactorial inheritance
Infection
Vascular accident
Toxins
Degenerative disease

per-second spike and wave pattern on the EEG. Inheritance of this EEG pattern seems to be an autosomal dominant trait but not all with the abnormal EEG develop petit mal.

Convulsions associated with fever (febrile fits) affect 1 in 20 children and the risk for sibs and offspring of an affected individual is increased to 1 in 5-10.

Affective psychosis

Affective psychosis includes cyclical depression and/or mania. One in 20 of the population is mildly affected and in 1% of the population the condition is severe. Twin and family studies support multifactorial inheritance with a risk to sibs and offspring of 1 in 6 for a proband with severe disease.

Schizophrenia

Schizophrenia is a serious psychiatric disorder which affects 1/100 of the population and accounts for about one half of all long term hospital inpatients in the UK. A mild disturbances of thought not amounting to full schizophrenia is called schizoid behaviour and this overlaps with behaviour which is considered within normal limits. Schizophrenia is subdivided into several clinical types but for each the risk to sibs and offspring is about 1 in 10.

Dementia

Progressive irreversible loss of higher cortical functions is heterogeneous. Identifiable causes include alcohol abuse, B12 deficiency and brain tumours. If onset occurs less than 65 years of age then this is termed presenile dementia. One in 20 of the population has presenile dementia and the frequency rises rapidly in older age groups (senile dementia). Where no cause can be identified the risk to relatives depends upon age of onset. For senile dementia the risks are probably not increased but presenile dementia may be autosomal dominant in some families.

Ageing

Ageing is universal yet it does not occur at a uniform rate. These individual differences may be under genetic control but evidence to support this is lacking.

In tissue culture fetal fibroblasts undergo a finite number (about 50) of divisions before dying. Fibroblasts from older individuals undergo fewer and fewer divisions before dying and this suggests that this cell line at least has an inherent limited lifespan.

Chapter 17
Congenital Malformations

Incidence

A malformation is an abnormality of structure due to defective embryogenesis. All malformations are thus congenital or present at birth, although they may not be diagnosed until later. Malformations may be single or multiple and may be of minor or major clinical significance. About 14% of newborns have a single minor malformation, 3% of newborns have a single major malformation and 0.7% of newborns have multiple major malformations (Table 17.1). The frequency of major malformations is even higher at conception (10-15%) but the majority of these fetuses are spontaneously aborted (Fig. 17.1). Thus 7-10% of spontaneous abortions are caused by major malformations.

Table 17.1 Classification and birth frequency of congenital malformations and deformations

Minor malformations	Single	140/1000
	Multiple	5/1000
Major malformations	Single	30/1000
	Multiple	7/1000
Deformations	Single	14/1000
	Multiple	6/1000

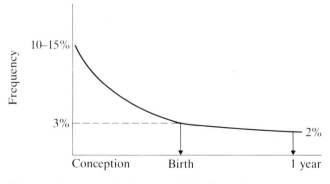

Fig. 17.1 Frequency of major congenital malformations

In contrast, a congenital deformation is an alteration in shape or structure of a previously normal part. These all arise after the period of embryogenesis but before birth. About 2% of newborns are affected with multiple deformations in one-third. Malformations and deformations may coexist and there is an increased risk of deformation (8%) in the presence of a major congenital malformation, especially of the central nervous system or urinary tract.

Aetiology
For 60% of all major congenital malformations the aetiology is obscure. Table 17.2 indicates the identifiable causes of congenital

Table 17.2 Aetiology of major congenital malformations

Idiopathic	60%
Multifactorial	20%
Monogenic	7.5%
Chromosomal	6%
Maternal illness	3%
Congenital infection	2%
Drugs, X-ray, alcohol	1.5%

malformations. Multifactorial inheritance is the commonest identifiable cause follwed by monogenic and chromosomal disorders. Thus genetic conditions account for at least one third of all congenital malformations or nearly 90% of congenital malformations of known aetiology.

There is a 6.4% risk of congenital malformation (especially congenital heart disease, neural tube defect and sacral agenesis) for the offspring of a diabetic mother and the risk is also increased for a mother with epilepsy although here it is difficult to separate the risk due to the disease and that due to her medications.

Certain drugs, for example thalidomide, phenytoin and warfarin are known to be teratogenic (Table 17.10). A few including aspirin, paracetamol, cephalosporins and aminoglycosides are not teratogenic but for the majority their safety in pregnancy is unknown and where possible should be avoided.

All of these identifiable agents must act during the critical period of active organogenesis.

CONGENITAL DEFORMATIONS

Deformations are caused by any factor which restricts the mobility of the fetus and so causes prolonged compression in an abnormal posture. Causes may be intrinsic or extrinsic (Table 17.3). Deformations are correctible by pressure. Complete resolution is usual in the newborn period (Table 17.4).

Table 17.3 Causes of congenital deformation

Intrinsic — neuromuscular disease, connnective tissue defects, CNS malformations
Extrinsic — primigravidae, small maternal stature, oligohydramnios, breech presentation, uterine malformation, multiple pregnancy

Table 17.4 Types of congenital deformation

Talipes
Congenital dislocation of the hip
Congenital postural scoliosis
Plaigocephaly
Torticollis
Mandibular asymmetry

Congenital dislocation of the hip

Four to 7 per 1000 livebirths have temporarily unstable or clicking hips; actual dislocation is found in 1/1000. The sex ratio is 6 females to 1 male. The risk of recurrence for sibs is 1 in 20 and for offspring is 1 in 8.

Talipes (club foot)

Five per 1000 livebirths have talipes and this is severe in 1/1000. The sex ratio is two males to one female with a sib risk of 1 in 50 for a male proband but 1 in 20 for a female proband. For an affected parent of either sex the risk of recurrence is 1 in 33.

MINOR CONGENITAL MALFORMATIONS

Table 17.5 lists some common minor congenital malformations. These are of no functional significance but should alert the clinician

Table 17.5 Minor congenital malformations

Epicanthic folds
Mongoloid or antimongoloid slant
Coloboma
Ear tag or pit
Bifid uvula
Simian crease
Fifth finger clinodactyly
Soft tissue syndactyly
Mongolian spot
Haemangioma
Umbilical hernia
Minor hypospadias
Single umbilical artery

to the possibility of an associated major malformation which coexists in over 90% of infants with multiple minor malformations.

MAJOR CONGENITAL MALFORMATIONS

Single major congenital malformations are present in 30/1000 neonates and Table 17.6 indicates the relative frequencies for each organ.

Table 17.6 Birth incidence of major congenital malformations

Organ	Incidence /1000 births
Brain	10
Heart	8
Kidney	4
Limbs	2
Other	6
TOTAL	30/1000

CENTRAL NERVOUS SYSTEM

The neural groove appears at 20 days and is mostly closed by 23 days. The anterior neuropore closes at 24 days and the posterior neuropore at 28 days. Although the most active period of embryogenesis is from the 3rd to 12th weeks, neuronal organisation and myelination continues throughout the whole pregnancy and up to 2-3 years of age.

Holoprosencephaly

In holoprosencephaly there is failure of the development of the forebrain and associated mid-face. This results in hypotelorism, bilateral cleft lip with an absent philtrum and severe mental retardation. In the most severe cases only a single central eye is present (cyclops). Death within 6 months is usual.

One half of cases are due to trisomy 13 but the rest are unexplained with a 6% empiric recurrence risk for siblings.

Isolated hydrocephalus

Hydrocephalus is an enlarged head due to interference with the normal circulation of cerebrospinal fluid. With succesful surgical shunting 80% have normal intelligence.

One per 1000 neonates has hydrocephalus and most of these are multifactorial with a 1.5% recurrence risk for siblings. Less than 1% of cases are due to an X-linked recessive trait and these boys have characteristic hypoplastic flexed thumbs (Fig. 17.2).

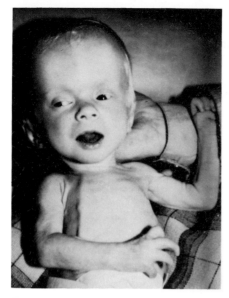

Fig. 17.2 X-linked hydrocephalus. Note characteristic thumb.

Prenatal diagnosis of hydrocephalus is often but not always possible in the second trimester by serial ultrasound scanning.

Microcephaly

Microcephaly is an abnormally small head as a result of defective brain growth. Identifiable causes include congenital infection, birth trauma, chromosomal imbalance maternal phenylketonuria and an autosomal recessive trait. Homozygotes for the autosomal recessive trait have a sharply slanted forehead whereas in microcephaly of other aetiologies the head is small but normally shaped (Fig. 17.3). Recurrence risk depends upon aetiology.

Prenatal diagnosis may be possible in a future pregnancy by serial scanning but sometimes the delay in head growth is not apparent until after the legal limit for termination of pregnancy.

Neural tube defect

Defective closure of the neural tube may occur at any level. Failure at the cephalic end (anterior neuropore) produces anencephaly and lower down produces spina bifida.

In anencephalics the cranial vault is missing and the exposed nervous tissue degenerates. The hypothalamus is defective and this leads to fetal adrenal atrophy with low maternal oestriols. Polyhydramnios may complicate the pregnancy. Stillbirth or neonatal death is invariable.

Spina bifida is most commonly lumbosacral with paralysis of the

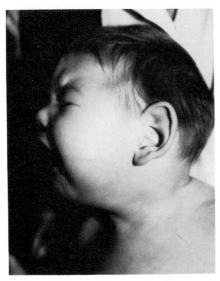

Fig. 17.3 Microcephaly secondary to congenital cytomegalovirus infection

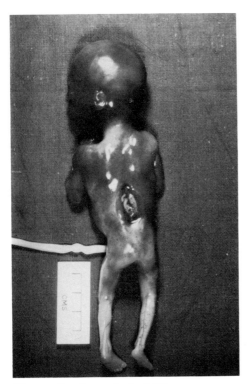

Fig. 17.4 Lumbosacral spina bifida

legs and sphincters (Fig. 17.4). About 15–20% have a covering of
intact skin (closed lesions) and these tend to cause less neurological
disability than the open lesions. Hydrocephalus commonly coexists.

The frequency of neural tube defects shows marked geographical variation. In USA, Canada, Japan, Africa and Mongolia 1/1000 births are affected. In South-East England the rate is 3/1000 and this rises to 5/1000 in the West of Scotland, 7.6/1000 in South Wales and 8.6/1000 in Northern Ireland. Eire has the highest incidence in the World of 10/1000. There appears to be an increased frequency in lower social classes and with winter-born children.

Neural tube defects also occur in 4.9% of all abortuses and so about 16/1000 of all conceptions are affected. Spina bifida and anencephaly have equal incidences and the two coexist in 20%.

In the UK the recurrence risk after an affected child is 1 in 25. The same recurrence risk applies to the offspring of an affected parent. For second degree relatives the risk is 1 in 50 and for third degree relatives it is 1 in 100. After two or more affected children the recurrence risk rises to 1 in 10. The recurrence risk is lower in countries with a lower incidence and in general is about ten times the incidence. Prenatal diagnosis is possible by a combination of ultrasound scanning and measurement of amniotic fluid alphafetoprotein. This will detect 100% of anencephaly and 98% of open spina bifida. In view of the frequency of this serious condition screening by assay of maternal serum alphafetoprotein is practised in high risk populations (Chapter 19).

HEART

The period of active cardiac organogenesis extends from the 3rd to the 8th weeks.

Many different congenital heart lesions have been described and the commoner are outlined in Table 17.7 together with the recurrence risks for family members. Overall congenital heart

Table 17.7 Incidence and recurrence risks for various types of congenital heart disease

Lesion	Birth incidence	Recurrence risk	
		for sibs	for offspring
Ventricular septal defect	1/400	1/25	1/25
Atrial septal defect	1/1000	1/33	1/33
Patent ductus arteriosus	1/830	1/33	1/25
Coarctation of aorta	1/1600	1/50	1/50
Aortic stenosis	1/2000	1/50	1/33
Common atrioventricular canal	1/2500	1/50	?
Total anomalous pulmonary venous drainage	1/5000	?	?
Transposition of the great vessels	1/16,000	1/50	?
Tetralogy of Fallot	1/1000	1/33	1/25

lesions affect 8/1000 births with a risk to sibs or offspring of about 1 in 25 and a risk to second degree relatives of less than 1 in 100. Recurrence is not specific to the type of heart lesion. The prognosis varies with the lesion and improvements in cardiac surgery have meant that some previously lethal lesions are now curable.

Prenatal diagnosis is possible for the more serious of these defects by detailed fetal echocardiology in specialised centres at 18-20 weeks of gestation.

GASTROINTESTINAL TRACT

The period of active organogenesis extends from 3 to 8 weeks of pregnancy. The intestinal loops have usually returned to the abdomen by the 12th week.

Anterior abdominal wall defects

Failure of the intestines to return to the abdomen occurs in 1/6000 pregnancies. Three types of lesion are identified (Table 17.8; Figs. 17.5-17.7). Exomphalos and gastroschisis may be surgically correctible in the absence of chromosomal abnormality or other malformations. Prenatal diagnosis by ultrasound is possible and as

Table 17.8 Classification of anterior abdominal wall defects

Type	Comments
Exomphalos	umbilical cord attached to apex of sac; sac may contain liver and/or intestines; chromosomal abnormality in 30%
Gastroschisis	no sac and umbilical cord not involved in defect; may be associated areas of atretic intestine and congenital heart disease (20%)
Body stalk anomaly	very short umbilical cord attached to apex of sac; severe spinal deformity; cloacal exstrophy; hypoplastic legs

these malformations may cause elevation of amniotic fluid and maternal serum alphafetoprotein they may be detected by a screening programme for neural tube defects (Chapter 19).

Cleft lip and/or palate

The lip is usually fused by day 35. If this fails it may impair closure of the palatal shelves which occurs at 8-9 weeks.

Cleft lip and/or palate occurs in 1/500 births and is most often due to multifactorial inheritance although it may be a feature of some single gene traits (eg. Treacher Collins syndrome) or of chromosomal abnormalities (eg. trisomy 13). Surgical repair without sequelae is usual.

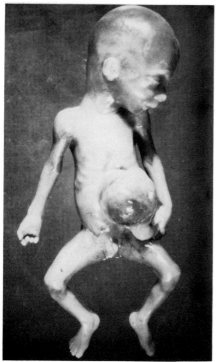

Fig. 17.5 Exomphalos   **Fig. 17.6** Gastroschisis

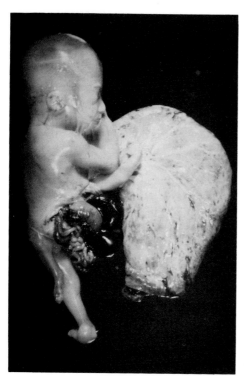

Fig. 17.7 Body stalk anomaly

For the multifactorial lesion the recurrence risk for normal parents with a child who has a unilateral cleft lip alone is 1 in 50 which rises to 1 in 20 if the child has bilateral cleft lip and palate. For second degree relatives the risk is 1 in 150 and for third degree relatives the risk is 1 in 300.

Cleft palate without cleft lip also shows multifactorial inheritance but is distinct from cleft lip and/or palate. Isolated cleft palate affects 1/2500 births with a recurrence risk of 1 in 50 for sibs and offspring.

Hirschsprung disease (Colonic aganglionosis)

Rectal biopsy confirms the absence of submucosal and myenteric ganglion cells which results in neonatal constipation and abdominal distension. A short or long segment may be involved, starting from the rectum.

One in 8000 neonates is affected with a 3 to 1 male excess. The condition is inherited as a multifactorial trait. Overall, for a male proband the risk to sibs is 1 in 50 but less than 1 in 100 to offspring. For a female proband the risk to sibs is 1 in 12 but again less than 1 in 100 for offspring. The risk also varies somewhat in proportion to the extent of the aganglionic segment.

Pyloric stenosis

Affected infants present in the first few weeks of life with vomiting and a palpably hypertrophic pylorus. The incidence in males is 1 in 200 with a 1 in 50 risk to sibs and a 1 in 25 risk to offspring. The incidence in females is 1 in 1000 with a 1 in 10 risk to sibs and a 1 in 5 risk to offspring.

Intestinal atresia

Atresia may occur at any level of the intestine with resulting symptoms and signs of obstruction. One in 330 neonates is affected. The lesion is sporadic with no increased recurrence risk.

Oesophageal atresia

Oesophageal atresia and/or tracheo-oesophageal fistula affects 1/3000 livebirths and like intestinal atresia is a sporadic event.

KIDNEY

The period of active renal organogenesis extends from the 4th to the 7th weeks. The overall frequency of renal and urinary tract malformations is 4/1000.

Bilateral renal agenesis

Absence of the kidneys results in oligohydramnios which in turn produces Potter's syndrome. The facies are squashed with large, low-set ears; other deformations are found; and pulmonary hypoplasia is usual (Figs. 17.8 and 17.9). Neonatal death is invariable.

The incidence is 1/3000 births and the empiric recurrence risk for the parents of an affected child is 1 in 33. Prenatal diagnosis is possible with ultrasound.

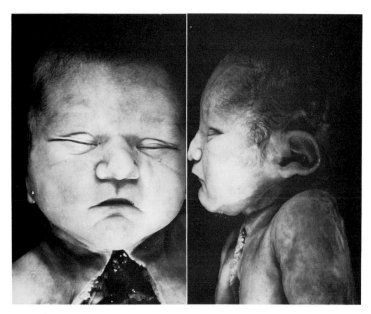

Fig. 17.8 and **17.9** Facies in bilateral renal agenesis (Potter's syndrome)

Renal hypoplasia and dysplasia

Congenitally small disorganised kidneys which may result in renal failure are aetiologically heterogeneous. A few cases are monogenic (e.g. Branchio-oto-renal syndrome) some are part of a multiple malformation syndrome (e.g. Noonan syndrome, Klippel- Feil syndrome or Turner's syndrome) but the majority are unexplained and have a low empiric recurrence risk.

Branchio-oto-renal syndrome (BOR syndrome)

The BOR syndrome is an autosomal dominant trait chracterised by preauricular pits, childhood deafness (of variable type) and renal hypoplasia (resulting in chronic renal failure in 6%). The frequency is 1 in 40,000.

Infantile polycystic disease

This autosomal recessive trait produces cysts in the liver, kidneys and pancreas which interfere with function and so result in death in early childhood. Prenatal diagnosis may be possible with ultrasound.

Adult polycystic disease

This common autosomal dominant trait affects 1/500 of the population. The renal cysts are asymptomatic until renal failure or hypertension ensue in the fourth decade. This condition accounts for 7% of all adults with chronic renal failure.

Offspring of an affected person can be screened with ultrasound but until the age of 30 years the disease cannot be confidently excluded. Prenatal diagnosis is not possible.

LIMBS

The period of active limb formation extends from 4 to 7 weeks of pregnancy. Overall major limb malformations are found in 2/1000 newborns. Table 17.9 lists the commoner of these together with their mode of inheritance.

Table 17.9 Types of limb malformation

Type	Birth incidence	Inheritance
Polydactyly	1/2000 (Caucasians)	usually sporadic
	1/200 (US negroes)	usually autosomal dominant
Syndactyly	1/1000	heterogeneous
Transverse limb defect	1/5000	sporadic
Ectrodactyly (split hand)	1/90,000	usually autosomal dominant but heterogeneous
Phocomelia (seal-like)	1/1000,000	usually sporadic but heterogeneous

THYROID

The thyroid gland develops from an outpouching of the floor of the pharynx during weeks 3-4 which descends into the neck and reaches its final postion by the 7th week and begins to function by the end of the 3rd month.

Thyroid dysgenesis

The thyroid is absent or severly hypoplastic in 1/3500 livebirths. The sex ratio is 4 females to 1 male. Birth weight and length are normal and symptoms are vague and not invariably present in the first few months of life (Table 19.5). Without therapy mental retardation and extreme short stature occur so now this condition is included in neonatal screening programmes (Chapter 19).

The defect is sporadic with no increased recurrence risk.

MULTIPLE CONGENITAL MALFORMATIONS AND DYSMORPHIC SYNDROMES

Multiple malformations, often a combination of minor and major, are present in 0.7% of newborns. In the majority the combination is random but for some the pattern is non-random and a dysmorphic syndrome can be identified. This distinction is important as accurate recurrence risks and prenatal diagnosis can be provided for many dysmorphic syndromes. Undoubtedly some of the idiopathic dysmorphic syndromes and cases of random multiple congenital malformations are due to submicroscopic chromosomal lesions since they share so many features with known chromosomal syndromes. In the following section some dysmorphic syndromes not covered in Chapter 14 (chromosomal disorders) or Chapter 15 (single gene disorders) are outlined.

Beckwith-Wiedemann syndrome

The clinical features are: macroglossia (90%), anterior abdominal wall defect (90%), high birth weight, ear lobe grooves and hemihypertrophy (15%) (Fig. 17.10). Neonatal hypoglycaemia may occur and result in mental retardation if untreated. Neoplasia ,

Fig. 17.10 Beckwith-Wiedemann syndrome. Note macroglossia and ear lobe grooves

especially Wilms tumour or adrenal cortical carcinoma occurs in 5%. One in 13,700 births is affected and it may be aetiologically heterogeneous as a small duplication of chromosome 11p15 has been found in some patients. For normal parents the sib recurrence risk is low.

CHARGE association

This is a sporadic non-random association of malformations: coloboma (80%), heart defects, choanal atresia (58%), retarded growth and development (87%), genital abnormalities (78%) and abnormal ears (88%). The name is an acronym from the features.

de Lange syndrome (Amsterdam dwarfism)

One in 30,000 births are affected by this condition which has an empiric sib recurrence risk of 1 in 50. Prenatal diagnosis is not possible.

Clinical features include: severe mental retardation, growth retardation, limb malformations, congenital heart disease (29%), cleft palate (20%), and a characteristic facies with thin lips, synophrys and anteverted nostrils (Fig.17.11). Death in early childhood is usual.

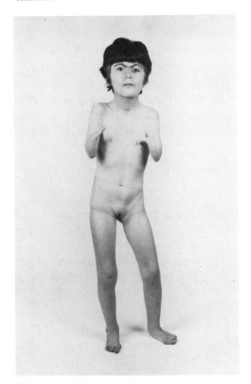

Fig. 17.11 de Lange syndrome

HARD ± E syndrome

The name of this autosomal recessive trait is an acronym from the clinical features: hydrocephalus, agyria (with mental retardation), retinal dysplasia and encephalocele (inconstant).

Klippel-Feil syndrome

This is characterised by a short stiff neck with a low hairline secondary to malformed cervical vertebrae. Other malformations may coexist: congenital heart disease (25%), renal malformations (30%), and Sprengel's shoulder.

This is a sporadic condition with an incidence of 1 in 42,000.

Noonan syndrome (Male Turner syndrome)

The Noonan syndrome is characterised by proportionate short stature (72%), variable hypogonadism, mild-to-moderate mental retardation (61%), congenital heart disease especially pulmonary stenosis (55%), low set ears, hypertelorism, ptosis, neck webbing, chest deformity, cubitus valgus and urinary tract malformations (27%).

The genetics of the Noonan syndrome are uncertain at least in part due to diagnostic difficulties with mild cases. For normal parents the sib recurrence risk is low.

Sturge-Weber syndrome

This sporadic syndrome results in a facial haemangioma wih a trigeminal distribution and an ipsilateral pia-arachnoid haemangioma. Glaucoma, epilepsy, hemiplegia and mental retardation may supervene.

VATER association

This is a sporadic non-random association of malformations: verterbral defects (70%); anal atresia (80%); tracheo-oesophageal fistula (70%), renal defects, radial limb dysplasia (65%), congenital heart disease and a single umbilical artery (35%). The name is an acronym from the involved organs. Intelligence is normal.

Williams syndrome

This sporadic syndrome affects 1/20,000. Features are: short stature, mental retardation, transient hypercalcaemia, supravalvular aortic stenosis and a characteristic facies with prominent lips and an anteverted small nose (Fig.17.12).

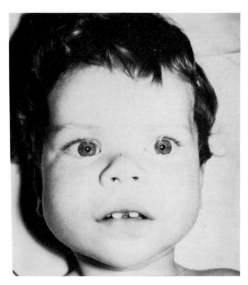

Fig. 17.12 Williams syndrome

MULTIPLE MALFORMATIONS DUE TO KNOWN TERATOGENS

Several environmental factors, called teratogens, are known which can cause malformations (Table 17.10). Numerous other agents are suspected to have teratogenic activity. The difficulty lies in establishing the causal relationship since animal experiments may not be directly informative. For example, thalidomide is teratogenic in rabbits but not in rats.

Table 17.10 Recognised human teratogens

Teratogen	Critical period	Malformations
Rubella	50% affected if infection in 1st 4 weeks; 0% > 12 weeks	Congenital heart disease, cataracts, microcephaly, mental retardation, sensorineural deafness
Cytomegalovirus	3rd or 4th month	Mental retardation, microcephaly
Toxoplasmosis	?1st trimester	Mental retardation, microcephaly
Alcohol	?1st trimester	Mental retardation, microcephaly, congenital heart disease, renal anomaly
Hydantoin	1st trimester	Hypoplasia of distal phalanges, cleft lip and palate, mental retardation
Thalidomide	34–50 days from LMP	Phocomelia, congenital heart disease, anal stenosis, atresia of external auditory meatus
Warfarin	1st trimester	Hypoplastic nose, stippled epiphyses, short distal phalanges, mental retardation

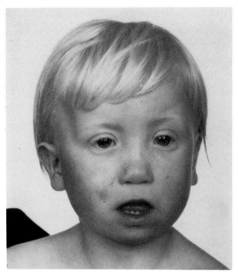

Fig. 17.13 Facies in the fetal alcohol syndrome

Malformations due to fetal infections with rubella, cytomegalovirus or toxoplasmosis are now uncommon. Each produces a distinctive pattern of malformations and often there is evidence of active neonatal infection with jaundice, purpura and hepatosplenomegaly. The diagnosis is confirmed by the demonstration in the neonate of raised levels of specific antibodies, especially IgM. Maternal immunity prevents recurrence in a future pregnancy. If a pregnant woman has evidence of seroconversion during the critical period then fetoscopic blood sampling to demonstrate evidence of fetal infection is of value in deciding whether termination of pregnancy is indicated.

Maternal alcohol ingestion of more than 150g per day poses a substantial risk to the fetus but lesser levels of intake may also be harmful. The facies in the fetal alcohol syndrome is characteristic with short palpebral fissures and a smooth philtrum (Fig. 17.13).

For all of these teratogenic agents a critical period has been identified outside which no malformation is produced. There also appear to be individual differences in susceptibility to these agents. For example, only 10% of offspring of mothers on hydantoin and only 25% of offspring of mothers on warfarin are affected. This susceptibility may reflect differences in fetal or maternal metabolism of the teratogen.

SUMMARY

Major congenital malformations are an important cause of fetal and neonatal morbidity and mortality. They are aetiologically heterogeneous. For example, congenital heart disease may be due to multifactorial inheritance, a single gene disorder (Holt-Oram syndrome), chromosomal imbalance, a teratogen (rubella) or an idiopathic dysmorphic syndrome (de Lange syndrome). This heterogeneity is of major importance when providing genetic counselling for affected individuals and their families.

Chapter 18
Prenatal Diagnosis

Prenatal diagnosis includes all aspects of embryonic and fetal diagnosis. Prenatal diagnosis is presently indicated in about 8% of all pregnancies and for these couples at increased risk of serious genetic disease it provides the reassurance without which many would decline to undertake a pregnancy. In practice 93% of prenatal tests provide reassurance for the couple concerned and selective termination of pregnancy is necessary in only 7%.

Prenatal diagnostic techniques may be divided in two broad groups: invasive and non-invasive (Table 18.1).

AMNIOCENTESIS

Amniocentesis is the withdrawal of amniotic fluid. This is usually performed at 16-18 weeks of gestation when there is about 180 ml of liquor and the ratio of viable to non-viable cells is maximal. Under aseptic conditions and after prior placental localisation with ultrasound, a needle is introduced into the amniotic cavity via the maternal abdomen. Ten to 20 ml of liquor is withdrawn and this can be used for several different tests (Table 18.2).

Table 18.1 Techniques for prenatal diagnosis

Invasive	— amniocentesis
	— fetoscopy
	— chorion villous sampling
Non-invasive	— ultrasound
	— radiography

Table 18.2 Tests which require amniocentesis

Fetal sexing
Fetal karyotyping
Fetal enzyme assay
Amniotic fluid biochemistry
Fetal DNA diagnosis

Fetal sexing

This may be required for female carriers of serious X-linked disorders where termination of any male pregnancy is considered (eg. Duchenne muscular dystrophy) or as a preliminary step before fetoscopy for the prenatal diagnosis of haemophilia or fragile X associated mental retardation.

Visualisation of the Barr body will usually permit fetal sexing within 3 hours. Y fluorescence may be misleading in view of the fluorescent autosomal heteromorphisms.

Fetal karyotyping

Fetal karyotyping is indicated for a maternal age of 35 years or over; a previous child with aneuploidy; pregancies where one parent has a balanced structural chromosomal rearrangement; and to confirm the fetal sex in X-linked conditions.

The amniotic fluid cells are grown in culture and a result is available in 2-3 weeks. About 1.5% of samples fail to grow and this is especially likely if the fluid was heavily blood stained. In experienced laboratories contamination of the fetal amniotic cells with maternal cells is unlikely.

Fetal metabolic disease

Prenatal diagnosis is now possible for more than 70 inborn errors of metabolism and is indicated if both parents are carriers (Table 18.3). The amniotic fluid cells are grown in culture for 4-6 weeks in order to provide sufficient cells for the assay of the appropriate enzyme. The enzyme level in these cells is compared with data from known normal and homozygous deficient amniotic fluid cells and with fibroblasts from the proband and parents.

Table 18.3 Examples of prenatally diagnosable inborn errors of metabolism

Lipid metabolism
   Tay Sachs disease, Gaucher disease, Niemann-Pick, familial hypercholesterolaemia, adrenoleucodystrophy, metachromatic leucodystrophy

Mucopolysaccharidoses

Aminoacid metabolism
   Methylmalonic acidaemia, homocystinuria, cystinosis, maple syrup urine disease, arginosuccinicaciduria

Carbohydrate metabolism
   Galactosaemia, glycogen storage disease (some types)

Others
   Lesch-Nyhan syndrome, acute intermittent porphyria, adenosine deaminase deficiency, xeroderma pigmentosum

Amniotic fluid biochemistry

Assay of alphafetoprotein (AFP) in amniotic fluid is indicated if there is a risk of neural tube defect as indicated by a previously affected child or a raised maternal serum AFP (Chapter 19).

The level of 17 alpha-hydroxyprogesterone in amniotic fluid is measured in pregnancies at risk for 21-hydroxylase deficient adrenogenital syndrome. Two dimensional electrophoresis of glycosaminoglycans can be used for the prenatal diagnosis of some types of mucopolysaccharidosis. Assay of the isoenzymes of alkaline phosphatase and of gammaglutamyl transpeptidase is under evaluation for the prenatal diagnosis of cystic fibrosis.

Fetal DNA diagnosis

The current indications for fetal DNA diagnosis are listed in Table 18.4. Recent advances in this area have been rapid and no doubt many more single gene disorders may soon be diagnosable with these techniques.

Table 18.4 Fetal DNA diagnosis

Direct detection of the point mutation or deletion
    Sickle cell disease
    Alpha thalassaemia

Linkage to restriction fragment length polymorphism
    Beta thalassaemia
    Haemophilia A*
    Haemophilia B
    Huntington's chorea*
    Phenylketonuria*
    Osteogenesis imperfecta*
    Duchenne muscular dystrophy*
    Fragile X associated mental retardation*

---

*under development

DNA is extracted from amniotic cells either directly or after culture. The diagnosis may be direct demonstration of the molecular defect or be established with a linked restriction fragment length polymorphism (Chapter 12).

Risks of amniocentesis

Amniocentesis carries a small additional risk of miscarriage for the pregnancy. For any pregnancy at 16 weeks gestation there is a 2.5% chance of spontaneous miscarriage. In experienced centres following amniocentesis the fetal loss rate is 2.8% and so the added risk from the procedure is 0.3% or 3/1000. If the indication for amniocentesis is raised maternal serum alphafetoprotein then a spontaneous abortion rate of 7% is found since AFP is raised in many non-viable pregnancies. The added risk from the procedure is again 0.3%. Maternal risks are negligible.

FETOSCOPY

Fetoscopy is the endoscopic visualisation of the fetus. The optimal time for this procedure is 18-20 weeks of gestation and it is restricted to a few specialised centres. Because of the size of the instrument only a limited field of view is possible and entire fetal visualisation is not practicable. Fetoscopy also provides a means for fetal blood sampling and for fetal skin or liver biopsy. Table 18.5 lists the main indications for fetoscopy.

In experienced hands the risk of spontaneous abortion after fetoscopy is under 5%. Recurrent leakage of amniotic fluid is a problem in 4% and premature labour occurs in 7-8%.

Table 18.5 Indications for fetoscopy

Fetal inspection	
Fetal blood sampling	— Haemophilia A and B*, fragile X mental retardation, beta thalassaemia*, alpha-1-antitrypsin deficiency, SCID, fetal infections, suspected mosaicism, in utero transfusion for Rh isoimmunisation
Fetal skin biosy	— Lethal epidermolysis bullosa
Fetal liver biopsy	— Ornithine transcarbamylase deficiency

* in cases where DNA diagnosis not possible

CHORIONIC VILLUS SAMPLING

Sampling of chorionic villi from the fetus is performed at 8-12 weeks of gestation. The biopsy is taken transcervically under direct vision or ultrasound guidance. No anaesthesia is necessary.

The material may be used for the same tests as amniotic fluid cells and has the advantage or rapid diagnosis since the cells need not be cultured before analysis. Fetal chromosomal analysis is possible in 24 hours. The spontaneous abortion rate at 10 weeks of pregnancy is 7% and the added rate from the procedure appears to be low although it has yet to be defined.

ULTRASONOGRAPHY

Visualisation of the fetus by ultrasound is completely safe for both mother and fetus. A wide range of congenital malformations may be diagnosed by an experienced ultrasonographer (Table 18.6). Ultrasound is indicated for a pregancy at increased risk for any of these disorders.

Table 18.6 Congenital malformations which can be diagnosed by ultrasound

CNS	— anencephaly, spina bifida (90%), hydrocephalus*, microcephaly*, encephalocele
Limb	— severe short-limbed dwarfism, polydactyly, severe osteogenesis imperfecta
Heart	— severe congenital heart disease
Kidney	— renal agenesis, bladder outflow obstruction, infantile polycystic disease
GIT	— duodenal atresia, anterior abdominal wall defect

* may not be detectable in all cases before 28 weeks of pregnancy

RADIOGRAPHY

The fetus can be seen on X-ray from 10 weeks. Occasionally radiography is indicated for the prenatal diagnosis of a skeletal dysplasia. The optimal time for such an X-ray is 20 weeks.

FETAL THERAPY

The prenatal diagnostic techniques outlined above are mostly directed to the detection of serious fetal abnormality with a view to termination of pregnancy. A limited number of conditions may be amenable to therapy which if instituted whilst in utero would give a better prognosis than if delayed until after birth (Table 18.7). So far

Table 18.7 Examples of fetal therapy

Blood transfusion — Rhesus isoimmunisation
Surgical bypass (fetoscopic cannulation) — obstructive uropathy, hydrocephalus
Medical manipulation (maternal diet, enzyme cofactors) — methylmalonicaciduria

direct blood transfusion at fetoscopy for severe Rhesus haemolytic disease has been the most widely practised of these procedures but other developments seem likely.

Chapter 19
Population Screening

Population screening entails the testing of a whole population in order to detect those at risk for a genetic disease. This approach is not appropriate for all genetic diseases as certain principles need to be observed (Table 19.1). Thus although many genetic diseases are well defined they are too rare to merit a whole population screening programme. There must be an advantage to early diagnosis for example carrier detection permits genetic counselling and perhaps prenatal diagnosis in a future pregnancy. Prenatal diagnosis provides the option of selective termination and neonatal diagnosis permits presymptomatic therapy. Any test used for population screening must be sensitive or cases will be missed and yet it must be relatively specific or else retesting of large numbers of false positives will be necessary.

Table 19.1 Principles of a screening programme

1. Clearly defined disorder
2. Appreciable frequency
3. Advantage to early diagnosis
4. Few false positives (specificity)
5. Few false negatives (sensitivity)
6. Benefits outweigh the costs

SCREENING FOR CARRIER DETECTION

At the present time in the UK there are no genetic diseases for which it is appropriate to screen the whole population for carriers. This may, however, change if carrier detection for cystic fibrosis became possible. Carrier screening is currently restricted to the haemoglobinopathies, Tay-Sachs disease and a few other conditions in ethnic groups at special risk (Table 19.2).

Beta thalassemia

The heterozygote frequency of beta thalassaemia varies widely in different populations but is especially high in Mediterranean

Table 19.2 Indications for carrier screening

Condition	Ethnic group(s)
Beta thalassaemia	Italians, Greeks, Cypriots, Indians, Turks, Thais, Chinese, Negroes
Alpha thalassaemia	Thais, Chinese, Negroes
Sickle cell disease	US and African negroes, West Indians
Tay-Sachs disease	Ashkenazi Jews

countries and South East Asia (Table 12.6, Chapter 12). Thus persons from these ethnic groups merit carrier screening. The carrier state is detected by the finding of microcytosis with a low mean cell haemoglobin (less than 25pg; normal range 27-30pg) and an increased hemoglobin A2 concentration (more than 3.5%; normal upper limit 2.5%). Carrier detection permits genetic counselling and prenatal diagnosis for couples who are both heterozygotes.

Sickle cell disease

Sickle cell disease is especially prevalent in Negroes (Table 12.3, Chapter 12). Heterozygotes are detected by the demonstration of sickled cells on exposure of red cells in vitro to a very low oxygen tension (Sickledex test). Detection of heterozygotes permits genetic counselling and alerts the anaesthetist prior to general anaesthesia.

Tay-Sachs disease

The heterozygote frequency for this autosomal recessive trait is 1 in 30 for Ashkenazi Jews but only 1 in 300 for other ethnic groups. Carriers are detected by the measurement of plasma hexosamini-dase A (a test which is unreliable during pregnancy). Carrier detection permits genetic counselling and prenatal diagnosis for couples at risk.

PRENATAL SCREENING

There are two major prenatal screening programmes at the present time in the UK: screening for fetal chromosome abnormalities in older mothers and alphafetoprotein screening for neural tube defects. Ultrasound screening for congenital malformations is as yet restricted to at risk pregnancies.

Chromosomal abnormalities

The frequency of trisomy 21 increases markedly with increasing maternal age (Fig. 14.2). Thus the risk at the 16th week of pregnancy to a 25 year old woman is 1 in 1000 whilst at 36 this risk is 1 in 200, at 39 it is 1 in 100 and at 42 years it is 1 in 50. Other aneuploidies also increase with increasing maternal age. All of these chromosomal abnormalities may be excluded by amniocentesis with fetal karyotyping.

The generally accepted recommendation is that amniocentesis should be offered to women who will be aged 35 years and over at their estimated date of confinement. In the UK there are 600-700,000 pregnancies each year and 4-5% of these (24-35,000) occur in women above this age limit. In practice there are about 14,000 amniocenteses for fetal chromosome analysis which is less than one half of those at risk.

Alphafetoprotein screening

Alphafetoprotein (AFP) is the major fetal plasma protein. AFP is made initially by the yolk sac and later by the liver. It peaks in the fetal bloodstream at 2-3 g/L around 12-14 weeks of gestation and falls thereafter. Neonatal blood levels fall rapidly but basal levels of 25 ng/l persist into adult life.

AFP is also found in the amniotic fluid at one hundredth the concentration found in fetal serum. Most of this is derived from fetal urine and again the maximum concentration of 50mg/L is attained at 14-15 weeks with a fall to below 10mg/L by 22 weeks of gestation. Fetal AFP also reaches the maternal blood stream and is detectable by radioimmunoassay. The maternal serum AFP rises from 13 weeks and peaks around 32-34 weeks at 500ng/ml before falling towards term.

In pregnancies where the fetus has a neural tube defect or certain other malformations, maternal serum AFP and amniotic fluid AFP are elevated by leakage from exposed fetal capillaries (Figs. 19.1 and 19.2). Measurement of maternal serum AFP is undertaken at 16-20

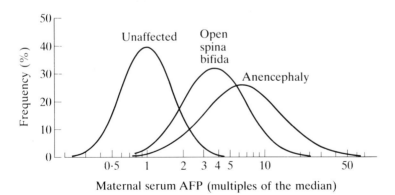

Fig. 19.1 Maternal serum AFP in normal pregnancies and pregnancies affected with neural tube defects

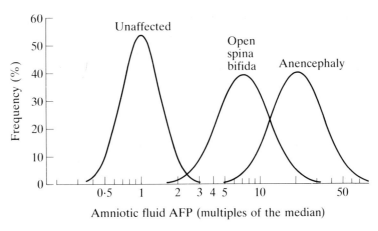

Fig. 19.2 Amniotic fluid AFP in normal pregnancies and pregnancies affected with neural tube defects

(optimum 17) weeks of gestation. If the level is above the 97th percentile a second blood sample is requested and ultrasound is advised to check the gestation and to exclude a missed abortion or a multiple pregnancy. If this sample also exceeds the 97th percentile then both amniocentesis for AFP measurement and detailed fetal ultrasound are indicated.

The causes of elevated AFP in maternal serum and amniotic fluid are listed in Table 19.3. In differentiating these causes it is useful to look at the pattern of cholinesterases in the amniotic fluid by polyacrylamide gel electrophoresis. Normal amniotic fluid produces only a single band of pseudocholinesterase. Open neural tube defects always have a faster second band of acetylcholinesterase as

Table 19.3 Causes of elevated maternal serum and amniotic fluid alphafetoprotein

Cause	Maternal serum AFP	Amniotic fluid AFP
Underestimated gestation	+	-
Overestimated gestation	-	+
Fetal blood in amniotic fluid	(+)	+
Multiple pregnancy	+	-
Threatened abortion	+	-
Anencephaly	+ +	+ +
Open spina bifida	+	+
Closed spina bifida	-	-
Isolated hydrocephalus	-	-
Anterior abdominal wall defect	+	+
Fetal teratoma	+/-	+/-
Maternal hereditary persistence of AFP	+ +	-
Congenital nephrotic syndrome	+	+
Skin defects	+	+
Placental haemangioma	+	+

do about 50% of anterior abdominal wall defects. Rarely a normal fetus may also have a second band and thus this test must be interpreted in conjunction with the amniotic AFP level and information from fetal ultrasound.

The amniocentesis rate with such a screening programme is less than 1% and the sensitivity of the test is 97% for anencephaly and 72% for spina bifida.

NEONATAL SCREENING

Neonatal screening was first introduced in 1961 for phenylketonuria. Its success has encouraged the development of other neonatal screening tests although the range of tests offered varies in different countries (Table 19.4). These tests are all performed a dried blood spot which is collected from a heel prick within the first two weeks of life (Guthrie card). In the UK virtually all newborns are tested.

Table 19.4 Conditions which may be included in neonatal screening programmes

Phenylketonuria
Galactosaemia
Congenital hypothyroidism
Congenital adrenal hyperplasia
Cystic fibrosis
Maple syrup urine disease
Homocystinuria
Duchenne muscular dystrophy

Phenylketonuria

Early diagnosis and therapy of phenylketonuria is mandatory if normal development is to occur. However, few if any physical signs are present in the neonate. The presence of an increased level of blood phenylalanine in the dried blood spot is detected by the Guthrie bacterial inhibition assay. Mild elevations of phenylalanine due to prematurity or delayed enzyme maturation are not uncommon and these can be excluded by repeat testing. False negatives are rare.

Congenital hypothyroidism

Early diagnosis and therapy will permit normal development yet few physical signs are present in the newborn (Table 19.5). Thyroid stimulating hormone (TSH) is measured on the dried blood spot. Neonates with primary hypothyroidism have elevated levels of TSH. The recall rate is 0.05% and false negatives appear to be rare.

Table 19.5 Signs in infants with congenital hypothyroidism at the time of diagnosis by newborn screening

Prolonged jaundice	80%
Open fontanelles	60%
Poor feeding	60%
Large tongue	47%
Hypothermia	40%
Umbilical hernia	35%
Hoarse cry	18%
Increased TSH	100%

PRESYMPTOMATIC SCREENING OF ADULTS

At the present time presymptomatic screening of adults is restricted to a few autosomal dominant conditions with delayed onset of symptoms (Table 19.6). This permits genetic counselling of

Table 19.6 Presymptomatic screening of adults

Condition	Screening test	Purpose
Polyposis coli	Sigmoidoscopy	Counselling and early colectomy
Familial hypercholesterolaemia	Lipid profile	Counselling and medical therapy
Huntington's chorea	Linked RFLP	Counselling
Dystrophia myotonica	EMG, ophthalmic examination	Counselling

affected individuals and may also be necessary for effective therapy; for example, colectomy prior to the development of cancer in polyposis coli. In the future identification of the 'at risk' genotype for the common chronic diseases of adulthood may be possible and screening to identify these individuals might permit avoidance of the environmental trigger(s) and so prevent disease.

Chapter 20
Prevention and Treatment of Genetic Disease

The old adage that prevention is better than cure applies equally to genetic as to acquired diseases. For diseases due to environmental agents such as infections the relationship between health and disease is outlined in Fig. 20.1. For genetic diseases the relationship is a little more complex since symptoms (i.e. disease) may not occur with an abnormal genotype until the person reaches adult life or until exposed to an environmental trigger. Primary prevention of

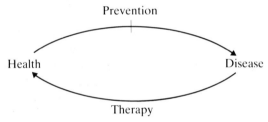

Fig. 20.1 Relationship between prevention and therapy for acquired disease

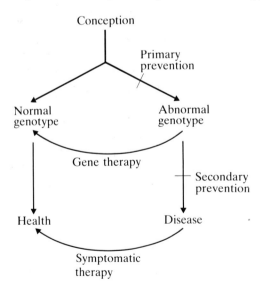

Fig. 20.2 Relationship between prevention and therapy for genetic disease

the abnormal genotype would need to act prior to conception (Fig. 20.2). Prenatal diagnosis with selective termination (secondary prevention) alters the birth frequency of the condition but is really a holding measure pending the development of primary prevention of genetic disease. If prevention fails then therapy is required.

OVERALL IMPACT OF GENETIC DISEASE

Genetic diseases affect all populations and have been apparent since prehistory. The recent focus of attention upon this group of conditions reflects the declining importance of infectious diseases as a cause of morbidity and mortality.

The infant mortality rate, IMR, (number of babies dying during the first year of life per 1000 livebirths) exemplifies this trend. The IMR for England and Wales was 154/1000 in 1900 with 4.5/1000 of the total due to genetic disease. By 1980 the IMR had fallen to 12/1000 due largely to improved public health and the control of infectious disease but the number due to genetic disease was unaltered. Thus the genetic contribution to the IMR has risen from 3% to 40%. The same pattern is apparent for fetal wastage and also for diseases of later childhood and adult life.

At least 7.5% of all conceptions have a chromosomal abnormality but the majority of these are lost as early spontaneous miscarriages. These early miscarriages also have a much higher rate (10-15%) of major congenital malformations than seen in the newborn (3%) with genetic causes accounting for at least one third of all malformations. Nearly one half of stillbirths and 20% of early neonatal deaths are attributabe to major congenital malformation. Genetic disease now causes about one half of all deaths in childhood and accounts for one third of all paediatric hospital admissions. Between 0.3% and 0.4% of children are severely mentally retarded with mild mental retardation in a further 3% and significant physical handicap in 1-2%. Most of this handicap is genetic in aetiology. Finally chronic diseases with a significant genetic component affect about 10-20% of the adult population.

The overall costs of genetic diseases to individuals, families and society are incalculable. Families commonly experience unjustifiable guilt which may result in divorce and neglect of the normal siblings. For society the predicted lifetime cost in 1982 for institutional and medical care for a single patient with trisomy 21 was one third of a million pounds.

TYPES OF GENETIC DISEASE

As indicated in Table 20.1, genetic diseases may be subdivided into chromosomal, single gene and multifactorial. Numerically the

Table 20.1 Incidence of genetic disease

Type	Incidence	
	/1000 live births	/1000 conceptions
Chromosomal	6	75
Single gene — autosomal dominant	10	⎫
— autosomal recessive	2	⎬ 14
— X-linked recessive	2	⎭
Multifactorial — major congenital malformations	6*	30*
— chronic adult	50**	50**
TOTAL	76/1000	169/1000

* The estimated multifactorial contribution of 20% (excludes chromosomal and monogenic contribution)
** The estimated multifactorial contribution of 33% (excludes the monogenic contribution)

multifactorial conditions are clearly the most important as they account for about 20% of all congenital malformations and about one third of all chronic conditions of adult life.

TREATMENT OF GENETIC DISEASE

There is a popular misconception that genetic disease is always untreatable. In practice some symptomatic therapy is available for the vast majority of genetic diseases and in some therapy can effectively restore normal health inspite of the continued presence of the abnormal genotype. Actual correction of the abnormal genotype is as yet experimental (Chapter 2).

Chromosomal disorders

For some sex chromosomal disorders sex hormone replacement will be required to permit normal development of secondary sex characteristics but cannot restore fertility. Autosomal imbalance usually results in mental retardation and multiple congenital malformations for which only symptomatic therapy is possible with perhaps medication for associated epilepsy and surgery for some of the malformations.

Single gene disorders

Table 20.2 indicates some of the commoner single gene disorders for which effective therapy is available.

Table 20.2 Examples of single gene diseases with effective therapies

Disease	Therapy
Congenital adrenal hyperplasia	Hormone replacement
Phenylketonuria	Dietary restriction of phenylalanine
Galactosaemia	Dietary restriction of galactose
Haemophilia	Factor replacement
SCID	Marrow transplant
Cystinuria	High fluid intake, D-penicillamine
Polyposis coli	Colectomy
Agammaglobulinaemia	Immunoglobulin replacement
Beta thalassaemia	Marrow transplant
Methylmalonic aciduria	Vitamin B12 - enzyme cofactor
Adult polycystic disease	Renal transplant
Wilson's disease	D-penicillamine

Multifactorial disorders

Table 20.3 indicates some of the commoner multifactorial disorders for which effective therapy is available. Diabetes mellitus was probably the first inherited disease to be effectively treated.

Table 20.3 Multifactorial diseases with effective therapies

Disease	Therapy
Cleft lip and palate	Surgery
Pyloric stenosis	Surgery
Congenital heart disease	Surgery, medications
Hydrocephalus	Surgery, medication
Diabetes mellitus	Medication
Hypertension	Medication
Peptic ulcer	Medication, surgery
Epilepsy	Medication

SECONDARY PREVENTION OF GENETIC DISEASE

Secondary prevention includes all aspects of prenatal screening with selective termination of affected pregnancies.

Chromosomal disorders

In theory secondary prevention could abolish chromosomal diseases. In practice screening is restricted to older mothers and to other high risk groups (Chapter 19). If all mothers of 35 years and over had amniocentesis then this would reduce the incidence of chromosomal disease by 30%. Although only about 4- 5% of pregnant mothers are above this age their increased incidence of aneuploidy accounts for the excess. If a maternal age of 40 years was taken then the reduction in chromosomal disease would be only 10%. Overall in the UK about 14,000 amniocenteses are performed per year to exclude chromosomal abnormality. This represents 2% of all pregnancies in the UK and results in about 150 terminations of abnormal fetuses each year.

At the present time not all mothers at increased risk of fetal chromosomal abnormality have an amniocentesis. In the West of Scotland the commonest cause of lack of screening is failure to offer the test rather than late booking or patient opposition to termination of pregnancy (Table 20.4).

Table 20.4 Amniocentesis for maternal age in the West of Scotland

25% screened
75% not screened — late booking (16%)
— patient's wish (7%)
— not offered (77%)

Single gene disorders

Prenatal diagnosis is possible for about 8% of autosomal recessive conditions by biochemical studies of amniotic fluid cells or chorion villous material. Only about 120 pregnancies per year in the UK are at risk for one of these prenatally diagnosable inborn errors of metabolism but this will double if the prenatal test for cystic fibrosis under current evaluation is substantiated.

For most X-linked conditions fetal sexing with termination of all male fetuses is the only option but for some specific prenatal tests are possible.

As yet few autosomal dominant traits are prenatally diagnosable.

The cost effectiveness of screening for carriers with prenatal diagnosis for couples at risk is exemplified by beta thalassaemia amongst Cypriot immigrants in London. For this condition screening throughout the UK could be run each year for less than the projected cost of treating one homozygote for one year.

Multifactorial disorders

Secondary prevention is now possible for a wide range of congenital malformations by the combined use of maternal alphafetoprotein assay and ultrasound. With such a screening programme in the West of Scotland the birth frequency of neural tube defects has been reduced by over 70% (Table 20.5).

As yet secondary prevention is not possible for any adult multifactorial disorders.

Table 20.5 West of Scotland screening for neural tube defects

	1976	1977	1978	1979	1980	1981	1982
NTD terminations	28	60	74	82	85	110	85
NTD births	152	124	103	82	63	43	na
NTD birth frequency /1000	4.3	3.6	2.9	2.2	1.7	1.1	na

na - not yet available

PRIMARY PREVENTION OF GENETIC DISEASE

Most couples who seek genetic counselling make an appropriate decision from the information (Table 20.6). Thus genetic counsel-

Table 20.6 Reproductive decisions after genetic counselling (data from Great Ormond Street)

	More children	No more children
High risk (<1 in 10)	1/3*	2/3
Low risk (>1 in 10)	3/4	1/4

* mild disorder or perinatal lethal

ling itself is a factor in primary prevention for all types of genetic disease. Although its impact is real enough for the family concerned it will have but little influence on overall gene frequencies as most families are not known to be at special risk until a genetic disease has occurred. Systematic counselling for an autosomal dominant trait with a low mutation rate such as Huntington's chorea may be an exception to this. Other approaches therefore need to be developed.

Chromosomal disorders

As chromosomal aberrations arise by non-disjunction or chromosomal breakage, an increased understanding of these processes is the prerequisite for primary prevention.

Single gene disorders

Ultimately all single gene disorders are the result of mutation so again research needs to be directed towards defining the causes of human mutation.

Multifactorial disorders

The multifactorial disorders offer perhaps the greatest scope for primary prevention. The goal here is to identify the environmental

Table 20.7 Multivitamin prophylaxis of neural tube defects

	Periconceptional vitamins	Controls
Total pregnancies	459	529
NTD pregnancies	3	24
Recurrence rate	0.7%	4.7%

and genetic components. Then those with the 'at risk' genotype can avert the disease by avoidance of the environmental factor.

Thus for example a multivitamin preparation may help to avert neural tube defects. Two groups of women were compared one with and one without vitamin supplementation during the periconceptional period. These women had all had a previous child with a neural tube defect and thus a 1 in 25 recurrence risk was expected. As seen in Table 20.7 the unsupplemented group had the expected rate of recurrence whereas the supplemented group had only a 1 in 140 recurrence rate. Although the two groups were not matched, this trial suggested that some component of the multivitamin preparation may help to avert a neural tube defect in those at risk. A randomized trial is in progress to confirm this work and to identify the active component of the vitamin preparation.

SUMMARY

Table 20.8 summarises the needs in relation to the goal of primary prevention of genetic disease.

Table 20.8 Summary of primary prevention

Chromosomal	Reduce non-disjunction and chromosomal breakage, genetic counselling
Single gene	Reduce mutation rate, genetic counselling
Multifactorial	Environmental prophylaxis, genetic counselling

Appendix 1
Chi-Square Test of Significance

This test is used to give a measure of the significance of an observed deviation from the expected value. The significance is related to the size of the sample. For instance, if a $1:1$ ratio is expected in a test involving 6 individuals, an observed ratio of $4:2$ is not too surprising by chance alone. However, if the test involved 600 individuals, an observed ratio of $400:200$ would be unlikely if the $1:1$ ratio really applies.

The chi-square (χ^2) test is designed to assess the significance of the deviation from the expected in relation to the number of tests carried out, or the number of observations made. It has the virtue of reducing many different samples of different sizes and with different numerical deviations, to a common scale for comparison.

The formula is used as follows:

$$\chi^2 = \Sigma(O - E)^2/E$$

Where χ^2 = chi-square
Σ = sum of
O = observed value
E = expected value

The significance of the value of chi-square is obtained from reference tables (abbreviated version below). The table is entered at the appropriate degree of freedom which is one less than the number of classes being tested. The larger the value of chi-square the less likely it is that the deviation is due to chance alone. A probability that the deviation is due to chance of less than 0.05 is taken to mean that other factors are playing a part and that the deviation from the hypothesis is significant.

For example, in Chapter 6 the autosomal recessive hypothesis was tested for sibships with albinism (Fig. 6.8). After correcting for bias of ascertainment 40 (23%) of 174 offspring of carrier parents were affected. With the autosomal recessive hypothesis 1 in 4 or 43.5 of the 174 should have been affected. How likely is this small deviation from expected by chance alone?

$\chi^2 = (40 - 43.5)^2/43.5 = 0.28$ for the affected class
$\chi^2 = (134 - 130.5)^2/130.5 = 0.09$ for the unaffected class
Total $\chi^2 = 0.37$ (one degree of freedom)

Table A1.1 Chi-square values

Degrees of freedom	Probability						
	0.99	0.8	0.5	0.1	0.05	0.01	0.001
1	0.00002	0.064	0.45	2.71	3.84	6.635	10.83
2	0.02	0.446	1.39	4.61	5.99	9.21	12.81
3	0.115	1.005	2.37	6.25	7.81	11.34	16.27
4	0.297	1.649	3.36	7.78	9.49	13.28	18.47
5	0.554	2.343	4.35	9.24	11.07	15.09	20.52

From the table this value of chi-square has a chance probability of 0.5 to 0.8 and so the deviation is not statistically significant and the autosomal recessive hypothesis is supported.

Appendix 2
Applications of Bayes' Theorem

Bayes' theorem is used in genetic counselling to combine other information with pedigree data in the assessment of an individual's chance of being a carrier for an X-linked recessive trait or being at risk for a late-onset autosomal dominant trait.

Carrier risk for an X-linked recessive trait

Fig. A2.1 shows the pedigree of a family with Duchenne muscular dystrophy. In this family I 2 is an obligate carrier as she has two affected sons. Thus her daughter II 3 has a 1 in 2 chance of also being a carrier. If she is a carrier then one half of her sons would be affected. She has four normal sons and thus she is either a carrier who has been lucky or more likely she is not a carrier. Bayes' theorem combines this conditional information (the normal sons)

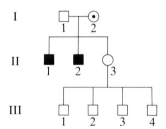

Fig. A2.1

Table A2.1

	II 3 carrier	II 3 not a carrier
Risk from pedigree (Fig. A2.1)	1/2	1/2
Conditional information (4 normal sons)	$(1/2)^4$	1^4
Joint (product of first two entries)	1/32	1/2 = 16/32
Final risk (each joint divided by the sum of the joints	1/17	16/17

Using the same denominator in the joint entry simplifies the last calculation as it becomes each numerator over the sum of the numerators.

Table A2.2

	II 3 carrier	II 3 not a carrier
Pedigree risk	1/2	1/2
Conditional information		
(4 normal sons and normal CPK)	$(1/2)^4 \times 1/3$	$1^4 \times 1$
Joint	1/96	48/96
Final	1/49	48/49

with the pedigree risk of 1 in 2 to produce a final modified risk (Table A2.1).

Thus the final risk that II 2 is a carrier is 1 in 17 a substantial reduction from the pedigree risk of 1 in 2.

A normal level of creatine phosphokinase (CPK) will also reduce the risk of II 3 being a carrier. For example, if her CPK was 50 i.u./l, then from Fig. 7.3 the relevant risk of being a carrier is 1 in 3. The conditional information (a normal level of CPK) is again combined with the pedigree risk using Bayes' theorem (Table A2.2). Her final risk of being a carrier is thus 1 in 49. In the near future, further conditional information from linked restriction fragment length polymorphisms may also be available to further modify the carrier risk.

Autosomal dominant traits with late onset

Fig. A2.2 shows the pedigree of a family with Huntington's chorea. In this family II 1 had a 1 in 2 chance of inheriting the mutant gene and the risk to III 1 is one half of this or 1 in 4. Onset is age dependent (Chapter 15) and at 70 years of age 95% of those with the

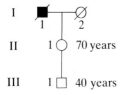

I

II   1   70 years

III   1   40 years

Fig. A2.2

Table A2.3

	II 1 has mutant allele	II 1 has normal allele
Prior risk (from pedigree)	1/2	1/2
Conditional information (no signs of		
Huntington's chorea at 70 years)	5/100	1
Joint	5/200	100/200
Final risk	1/21	20/21

mutant allele show the disease. II 1 is showing no signs of Huntington's chorea at 70 years of age and this conditional information can again be combined with the pedigree risk using Bayes' theorem (Table A2.3). Thus the final risk that II 1 has the mutant allele is 1 in 21 and the risk to III 1 is one half of this or 1 in 42. This risk will be modified still further as both he and his mother remain healthy.

These risks can be modified further with conditional information from a closely linked RFLP in certain families.

Appendix 3
Calculation of the coefficients of relationship and inbreeding

The coefficient of relationship (r) is the proportion of all genes in *two* individuals which are identical by descent. Calculation of this may be helpful in providing a recurrence risk for an autosomal recessive trait for members of an inbred family.

The coefficient is calculated from the formula:

$$r = (1/2)^n$$

where n is the number of steps apart on the pedigree for the two individuals via the common ancestor. If there is more than one common ancestor then their contributions are added to give a final r value

For example in Fig. A3.1 first cousins have an r value of:

$$r = (1/2)^4 + (1/2)^4 \text{ or } 1/8$$

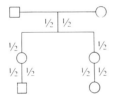

Fig. A3.1

Thus on average 1 in 8 of the genes of first cousins are identical by descent.

The coefficient of inbreeding (F) is the proportion of loci at which *one* individual is homozygous by descent. Thus if first cousins married their child's proportion of loci which would be homozygous by descent would be on average one half of the proportion of parental genes identical by descent or r/2.

Thus $F = r/2$

In Fig. A3.2 a man by his first wife has a child with an autosomal recessive disorder. He then marries his first cousin. What is the risk of recurrence?

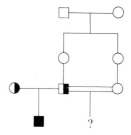

Fig. A3.2

He is an obligate heterozygote. Since r for first cousins is 1 in 8 then the chance that his wife has the same recessive allele from a common ancestor is 1/8. For two heterozygotes the risk of recurrence is 1 in 4. Thus the final risk is the product of these probabilities:

1 (his carrier risk) × 1/8 (her carrier risk) × 1/4 = 1/32

Appendix 4
Self-Assessment Section

The following questions approximately follow the order of the text. In each question any number of statements may be true or false. Answers are on page 265.

1. Human DNA

A. contains purines (adenine and thymine)
B. is double-stranded
C. is 3 million kilobases long in the haploid genome
D. contains large stretches which are apparently without function
E. codes for 50-100,000 structural genes

2. Human messenger RNA

A. may be single- or double-stranded
B. has uracil in place of thymine
C. is complementary to the transcribed gene
D. is translated 3' to 5'
E. requires processing before it is functional

3. Insulin

A. is encoded by a gene on chromosome 12
B. can be synthesised by genetic engineering
C. is 55 aminoacids long
D. is necessary in diabetes mellitus as the gene is deleted

4. Point mutations

A. may be visible with the light microscope
B. in 50% of cases lead to no aminoacid change due to code degeneracy
C. may be increased be ionising radiation
D. occur approximately once per million times a gene is replicated

254

5. Restriction endonucleases

A. are bacterial lysosomal enzymes
B. cleave DNA only at specific sequences
C. EcoRI produces flush ends
D. may be useful in prenatal diagnosis

6. Restriction fragment length polymorphisms

A. are only found on the X chromosome
B. cause no clinical abnormality
C. may be linked to a disease locus of interest
D. are routinely used to confirm zygosity

7. Chromosomes

A. are present in all human cells
B. are routinely stained with Giemsa banding
C. show polymorphisms in at least 30% of the population
D. 13,14,15, 21 and 22 have satellites
E. are also present in mitochondria

8. Mitosis

A. only occurs after puberty
B. takes approximately 24 hours in cultured mammalian cells
C. occurs immediately after G1
D. Sister chromatid exchanges may occur in mitosis

9. Meiosis

A. has five stages during the prophase of the first division
B. in the male results on average in 72 cross-overs
C. may produce more than 8 million different chromosomal combinations
D. takes about 75 days in the male
E. is arrested at diplotene in the female until puberty

10. Lyonisation

A. occurs at or before the 2000 cell stage in the human
B. results in inactivation of all genes on one X chromosome in female somatic cells
C. results in an inactive X which is slow to replicate
D. results in an inactive X which may be seen as the Barr body
E. may produce clinical symptoms in a female carrier of an X-linked recessive trait

11. Chromosome aberrations

A. affect about 7.5% of all conceptions
B. are found in 0.5% of still births
C. arise pricipally from non-disjunction and chromosomal breakage
D. 69,XYY is the commonest form of triploidy

12. Translocations

A. Lifespan is reduced in carriers of balanced translocations
B. may be reciprocal, centric fusion or insertional
C. Centric fusion translocations result in a trivalent at meiosis
D. Centric fusion of chromosomes 13 and 14 is the single commonest type of human translocation

13. Deletions

A. The smallest visible chromosomal loss is about 6 million base pairs
B. Autosomal deletions tend to be more serious than deletions of the sex chromosomes
C. may arise when one parent has a balanced centric fusion translocation
D. may arise when one parent has a pericentric inversion

14. Chromosomal aberrations

A. Mosaicism always arises after fertilization
B. Mosaicism may be confined to the gonad
C. A chimaera is derived from two separate zygotes
D. A hydatidiform mole is always 46,XY

15. Autosomal dominant traits

A. usually result in equal numbers of affected males and females
B. may be transmitted from males to males or females and vice versa
C. often show variable expression
D. may be non-penetrant
E. may show a maternal age effect for new mutations

16. Autosomal recessive traits

A. usually result in equal numbers of affected males and females
B. produce a horizontal pedigree
C. have an increased frequency in consanguineous families
D. may show genetic heterogeneity
E. include Tay-Sachs disease and congenital spherocytosis

17. X-linked recessive traits

A. result in a marked excess of affected males
B. are not transmitted from male to male
C. may manifest in carrier females due to atypical lyonisation
D. may occur in females with X-autosome translocations
E. have an increased frequency in consanguineous families

18. X-linked dominant traits

A. include incontinentia pigmenti and vitamin D-resistant rickets
B. are passed from a mother to all daughters but only to one half of her sons
C are more variable in affected females because of lyonisation
D. overall occur twice as often in females as males

19. Linkage

A. means that two loci are within measurable distance on the same chromosome
B. disturbs independent assortment
C. The Lod score gives a measure of the likelihood of a degree of disturbance of independent assortment by chance alone
D. A Lod score of 5 would indicate linkage of two loci
E. Linkage in coupling is more common than linkage in repulsion

20. Gene mapping

A. The loci for nail-patella syndrome and the ABO blood group are linked and occur on chromosome 8
B. The locus for beta thalassaemia is on chromosome 11
C. L1.28 is a DNA probe which hybridises to the locus for Duchenne muscular dystrophy
D. Autosomal cross-overs are more frequent in males than females
E. Acid phosphatase is determind by a gene on 22p

21. Gene mapping

A. The immunoglobulin kappa light chain gene is on chromosome 2
B. The insulin gene is on chromosome 11
C. Thymidine kinase was mapped to chromosome 15 by somatic cell hybridisation
D. Loci for enzymes in the same pathway usually occur on the same chromosome
E. About 10% of the loci for human single gene traits are now firmly assigned

22. Multifactorial inheritance

A. The pedigree pattern is usually diagnostic
B. include continuous and discontinuous traits
C. is investigated by twin concordance and family linkage studies
D. is of little practical importance

23. Twins

A. occur once in every 189 pregnancies
B. are identical if monochorionic
C. are identical if monoamniotic
D. Identical twins have a higher concordance than non-identical for all genetic traits

24. Continuous multifactorial traits

A. often show a Gaussian distribution
B. include height, intelligence and skin colour
C. arise from interaction of several loci with environmental factors
D. show regression to the mean

25. Discontinuous multifactorial traits

A. include many congenital malformations and common chronic adult disorders
B. occur when a threshold is exceeded
C. generally have lower recurrence risks than single gene traits
D. have an increased recurrence risk if the proband is of the less frequently affected sex

26. The Hardy-Weinberg law

A. demonstrates why dominant traits do not automatically increase in frequency
B. can be used to calculate mutation rates
C. shows that carrier frequencies are high even for very rare autosomal recessive traits
D. is disturbed by selection

27. Selection

A. is or was involved in the creation of polymorphisms
B. acts quickly when operating against the recessive homozygote
C. operated on the heterozygotes for sickle cell anaemia
D. may result in allele fixation

28. Genetic poymorphisms

A. have a rare allele frequency of 1 in 50 or greater
B. are found in about 3% of enzymes
C. include the ABO blood group and many restriction fragment length polymorphisms
D. are useful in linkage studies

29. Immunoglobulins

A. are exceptions to the one gene-one polypeptide rule
B. all have the same two types of light chain
C. are absent in SCID
D. are absent in DiGeorge syndrome
E. Isoagglutinins are predominantly IgG

30. Haemolytic disease of the newborn

A. is now always preventible
B. may arise when a Rh +ve mother has a Rh -ve fetus
C. may result in hydrops fetalis
D. may result in kernicterus

31. HLA

A. is a gene complex on chromosome 16
B. Certain HLA haplotypes cause diseases such as ankylosing spondylitis
C. is of little importance in marrow transplantation
D. The HLA antigens are not present on sperm or erythrocytes

32. Haemoglobin

A. HbA_2 is the major adult haemoglobin
B. Alpha globin chains are found in HbF
C. The alpha chain locus is on chromosome 11
D. The beta chain locus is on chromosome 16
E. More than 350 abnormal haemoglobins have been described

33. Sickle cell disease

A. is due to a beta globin point mutation
B. produces few ill effects in the homozygote
C. Homozygotes are very susceptible to pneumococcal infections
D. Prenatal diagnosis is possible by DNA analysis

34. Alpha thalassaemia

A. is usually due to gene deletion
B. HbH is due to loss of 3 of the 4 alpha genes
C. Carrier detection is straightforward
D. Prenatal diagnosis is possible but requires fetal blood sampling

35. Beta thalassaemia

A. is usually due to gene deletion
B. shows marked heterogeneity of molecular pathology
C. Carrier detection is straightforward
D. Prenatal diagnosis is possible but always requires fetal blood sampling

36. Oncogenes

A. are normal components of human DNA
B. may cause cancer if their normal regulation is disturbed
C. include c-abl and c-myc
D may be activated by translocation into the immunoglobulin gene complex

37. Indications for chromosomal analysis include:

A. recurrent miscarriages
B. Prader-Willi syndrome
C. multiple congenital malformations
D. Treacher Collins syndrome
E unexplained mental retardation

38. Pitfalls in genetic counselling include:

A. inadequate knowledge of the literature
B. genetic heterogeneity
C. non-penetrance
D. gonadal mosaicism
E. incorrect diagnosis

39. Chromosomal disorders

A. occur in 60% of early spontaneous abortions
B. Trisomy 16 is especially frequent in early spontaneous abortions
C. 45,X is especially frequent in early spontaneous abortions
D. Deletions tend to be more harmful than duplications
E. The birth frequency of balanced translocations is 1/1000

40. Trisomy 21

A. has an overall birth incidence of 1/700
B. shows a marked maternal age effect
C. always results in mental retardation (unless mosaic)
D. may be complicated by hypothyroidism

41. Klinefelter's syndrome

A. is found in 10% of infertile males
B. generally results in mental retardation
C. is usually due to paternal non-disjunction
D. has a 1 in 100 sib recurrence risk

42. Fragile X associated mental retardation

A. Affected males may have enlarged testes
B. 20-30% of carrier females show some degree of mental retardation
C. Prenatal diagnosis is currently not possible
D. Carrier detection is straightforward

43. Mental retardation

A. affects 3.4% of the population
B. has a trivial recurrence risk where the cause cannot be identified
C. is a complication of 45,X
D. is a complication of 47,XXX

44. Achondroplasia

A. is the commonest cause of short-trunked dwarfism
B. shows characteristic narrowing of the lumbar interpedicular distance
C. may cause spinal cord comression
D. is due to a mutant collagen gene on chromosome 13

45. Congenital adrenal hyperplasia (21-hydroxylase deficiency)

A. has a high prevalence in Apache Indians
B. Prenatal diagnosis is not possible
C. Carrier detection is not possible
D. The locus is on chromosome 6

46. Cystic fibrosis

A. affects 1/12,000 Caucasians
B. Carrier detection is not possible
C. is uncommon in Orientals
D. can be screened for in the newborn

47. Huntington's chorea

A. is due to a mutant gene on chromosome 4
B. has complete but age-dependent penetrance
C. has a high frequency in Tasmania
D. is incurable

48. Duchenne muscular dystrophy

A. is due to a mutant gene at Xp21
B. may be allelic to Becker muscular dystrophy
C. Prenatal diagnosis, other than by fetal sexing, is not yet possible
D. Carrier detection by CPK testing is not absolute

49. Phenylketonuria

A. Prenatal diagnosis is possible
B. Neonatal screening is possible
C. Carrier detection is not possible without DNA analysis
D. The locus is on chromosome 13

50. Glucose 6-phosphate dehydrogenase deficiency

A. The locus is on Xpter
B. More than 1500 variant forms are known
C. Carrier females are usually asymptomatic
D. Carbohydrates may precipitate a haemolytic crisis

51. Diabetes mellitus

A. is uncommon in Eskimoes
B. Maturity onset diabetes of youth is inherited as an autosomal dominant trait
C. Insulin independent diabetes mellitus affects about 10% of the the population
D. The sib recurrence risk after one affected child with insulin dependent diabetes mellitus is 1 in 33

52. Multiple sclerosis

A. affects 1/2000 individuals in the UK
B. has an HLA association with B27
C. has a sib and offspring risk of 1/100
D. Prenatal diagnosis is possible

53. Major congenital malformations

A. occur in 10-15% of conceptions
B. are most commonly unexplained
C. are increased in frequency in the offspring of mothers with diabetes mellitus or epilepsy
D. are always present at birth although they may not be diagnosed until later

54. Neural tube defects

A. occur before the 21st day of gestation
B. have a maximum prevalence in the West of Scotland
C. have a recurrence risk of 1 in 125
D. Prenatal diagnosis is possible

55. Anterior abdominal wall defects

A. are of three main types
B. Chromosomal aberration coexists in 10% with gastroschisis
C. Congenital heart disease is present in most with a body stalk anomaly
D. Prenatal diagnosis is possible

56. Bilateral renal agenesis

A. may result in oligohydramnios
B. occurs 1/30,000 births
C. has a recurrence risk of 1/33
D. Prenatal diagnosis is not possible

57. Fetal alcohol syndrome

A. may occur with paternal alcohol abuse
B. includes mental retardation
C. A smooth philtrum is characteristic
D. is a cause of neonatal hepatic cirrhosis

58. Amniocentesis is indicated for:

A. fetal sexing
B. pregnancy at risk for neurofibromatosis
C. pregnancy at risk for mucopolysaccharidosis
D. when one parent has a balanced pericentric inversion of chromosome 9
E. prenatal diagnosis of alpha thalassaemia

59. Chorionic villus sampling

A. is not possible after the 14th week of gestation
B. allows rapid fetal karyotyping
C. may be used for the prenatal diagnosis of beta thalassaemia
D. may be used for the prenatal diagnosis of anencephaly

60. AFP is usually elevated in maternal serum when the pregnancy is complicated by:

A. open neural tube defect
B. anterior abdominal wall defect
C. congenital nephrotic syndrome
D. placental haemangioma
E. isolated hydrocephalus

Chapter 2

1. A F
    B T
    C T
    D T
    E T

2. A F
    B T
    C T
    D F
    E T

3. A F
    B T
    C F
    D F

4. A F
    B F
    C T
    D F

5. A F
    B T
    C F
    D T

6. A F
    B T
    C T
    D F

Chapter 3

7. A F
    B T
    C T
    D T
    E T

8. A F
    B F
    C F
    D T

Chapter 4

9. A T
    B F
    C T
    D F
    E F

10. A T
    B F
    C T
    D T
    E T

Chapter 5

11. A T
    B F
    C T
    D F

12. A F
    B T
    C T
    D T

13. A T
    B T
    C F
    D T

14. A T
    B T
    C T
    D F

Chapter 6

15. A T
    B T
    C T
    D T
    E F

16. A T
    B T
    C T
    D T
    E F

Chapter 7

17. A T
    B T
    C T
    D T
    E F

18. A T
    B F
    C T
    D T

Chapter 8

19. A T
    B T
    C T
    D T
    E F

20. A F
    B T
    C F
    D F
    E F

21. A T
    B T
    C F
    D F
    E T

Chapter 9

22. A F
    B T
    C F
    D F

23. A F
    B T
    C T
    D T

24. A T
    B T
    C T
    D T

25. A T
    B T
    C T
    D T

Chapter 10

26. A T
    B F
    C T
    D T

27. A T
    B F
    C T
    D F

28. A T
    B F
    C T
    D T

Chapter 11

29. A T
    B T
    C T
    D F
    E F

30. A F
    B F
    C T
    D T

31. A F
    B F
    C F
    D T

Chapter 12

32. A F
    B T
    C F
    D F
    E T

33.	A	T	40.	A	T	48.	A	T	56.	A	T
	B	F		B	T		B	T		B	F
	C	T		C	T		C	T		C	T
	D	T		D	T		D	T		D	F

34.	A	T	41.	A	T	49.	A	T	57.	A	F
	B	T		B	F		B	T		B	T
	C	F		C	F		C	T		C	T
	D	F		D	F		D	F		D	F

35.	A	F	42.	A	T	**Chapter 16**			**Chapter 18**		
	B	T		B	T	50.	A	F	58.	A	T
	C	T		C	F		B	F		B	F
	D	F		D	F		C	T		C	T
							D	F		D	F
										E	T

36.	A	T	**Chapter 15**			51.	A	T	59.	A	T
	B	T	43.	A	T		B	T		B	T
	C	T		B	F		C	F		C	T
	D	T		C	F		D	T		D	F
				D	T						

Chapter 13			44.	A	F	52.	A	T	**Chapter 19**		
37.	A	T		B	T		B	F	60.	A	T
	B	T		C	T		C	T		B	T
	C	T		D	F		D	F		C	T
	D	F								D	T
	E	T				**Chapter 17**				E	F

45.	A	F	53.	A	T
	B	F		B	T
	C	F		C	F
	D	T		D	T

38.	A	T						
	B	T	46.	A	F	54.	A	F
	C	T		B	T		B	F
	D	T		C	T		C	F
	E	T		D	T		D	T

Chapter 14								
39.	A	T	47.	A	T	55.	A	T
	B	T		B	T		B	F
	C	T		C	T		C	F
	D	T		D	T		D	T
	E	F						

Glossary

Alleles – alternative forms of a gene at the same locus

Amniocentesis – aspiration of amniotic fluid

Aneuploid – any chromosome number which is not an exact multiple of the haploid number

Ascertainment – identification of families with an inherited condition

Autosome – any chromosome other than the sex chromosomes

Autozygous – homozygous at a locus for alleles identical by descent

Backcross – mating of a heterozygote with a recessive homozygote

Bacteriophage – a bacterial virus

Bivalent – a pair of homologous chromosomes as seen following synapsis prior to the first meiotic division

Burden – consultand's perception of the cost (emotional, physical and financial) of a genetic disorder

Carrier – a recessive heterozygote

Chiasma – the crossing of chromatid strands of homologous chromosomes during meiosis as a result of meiotic recombination

Chimaera – an individual whose cells are derived from more than one zygote

Chromatid – replicated DNA prior to separation at mitosis

Chromatin – the nucleoprotein fibre of a chromosome

Chromosomal aberration – any abnormality of chromosome number or structure visible under the microscope

Clone – a cell line derived by mitosis from a single diploid cell

Codominant – both alleles of a pair are expressed in the heterozygote

Codon – three adjacent bases in DNA or RNA which specify an aminoacid

Coefficient of inbreeding (F) – the proportion of loci at which an individual is homozygous by descent

Coefficient of relationship (r) – the proportion of all genes in two individuals which are identical by descent

Compound – an individual with two different mutant alleles at a locus

Concordant – both members of a twin pair show the trait

Congenital – present at birth

Consanguineous – mating between individuals who share at least one common ancestor

Consultand – any person requesting genetic counselling

Cosmid – synthetic cloning vector which can accomodate large fragments of foreign DNA

Crossover – exchange of genetic material between homologous chromosomes during meiosis

Diploid – the chromosome number of somatic cells

Discordant – only one member of a twin pair shows the trait

Dominant – a trait expressed in the heterozygote

Empiric risk – recurrence risk based on experience rather than calculation

Eukaryote – any organism with a nucleus and a nuclear membrane

Exon – a coding area of a gene

Expressivity – variation in the degree of gene expression

F₁ – the first generation progeny of a mating

Fetoscopy – endoscopic visualisation of the fetus

Forme fruste – an incomplete, partial or mild form of a trait or syndrome

Gene – a sequence of bases in DNA which codes for one polypeptide

Gene pool – all the genes present at a given locus in the population

Genetic counselling – the communication of information and advice about inherited disorders

Genetic engineering – the artificial production of new combinations of heritable material

Genetic lethal – a genetic disorder in which affected individuals fail to reproduce

Genetics – the scientific study of variation and heredity

Genome – the genetic constitution of an individual

Genotype – the alleles present at one locus

Haploid – the chromosome number of gametes

Hemizygous – genes on the X chromosome in males

Heredity – the transmission of characteristics to descendents

Heritability – a statistical measure of the degree to which a trait is genetically determined

Holandric – Y-linked inheritance

Hybrid – a species cross

Heterogeneity – similar phenotypes from different genotypes

Heterozygote – an individual with one normal and one mutant allele at a given locus on a pair of homologous chromosomes

Homologous – matched

Homeostasis – the tendency for the internal environment of the body to remain constant inspite of varying external conditions

Homozygote – an individual with a pair of identical alleles at a given locus on homologous chromosomes

Idiogram – a diagram of the chromosome complement

Inbreeding – the mating of closely related individuals

Intron – a non-coding area of a gene

Isochromosome – an abnormal chromosome with duplication of one arm and deletion of the other caused by transverse division of the centromere

Isolate – a genetically separate population

Isozymes – multiple molecular forms of an enzyme

Karyotype – the classified chromosome complement of an individual or cell

Kindred – an extended family

Linkage – linked genes have their loci within measurable distance of one another on the same chromosome

Locus – the precise location of a gene on a chromosome

Meiosis – reduction cell division which occurs in gamete production

Mitosis – somatic cell division

Mongrel – a variety cross

Monosomy – one of a chromosome pair is missing

Mosaic – an individual derived from a single zygote with cells of two or more different genotypes

Multifactorial – inheritance due to multiple genes at different loci which summate and interact with environmental factors

Mutation – a change in the genetic material

Nondisjunction – failure of two members of a chromosome pair to disjoin during anaphase

Nucleotide – a purine or pyrimidine base attached to a sugar and phosphate group

Oncogene – a gene sequence capable of causing transformation

Palindrome – a stretch of DNA in which identical base sequences run in opposite directions

Penetrance – the frequency of expression of the genotype

Phenocopy – an environmentally induced mimic of a genetic disease

Phenotype – the observable characteristics of an individual

Plasmid – extrachromosomal closed circular DNA molecule found in bacteria

Pleiotropy – multiple effects of a single gene

Polygenic – determined by multiple genes at different loci each with a small but additive effect

Polymorphism – the occurrence together in a population of two or more discontinuous traits in such proportions that the frequency of the rarest could not be maintained only by recurrent mutation

Polyploid – an abnormal chromosomal complement which exceeds the diploid number and is an exact multiple of the haploid number

Proband – the individual who draws medical attention to the family

Probe – a radiolabelled DNA fragment used to identify a complementary sequence(s)

Prokaryote – a simple unicellular organism which lacks a nuclear membrane

Race – a group of historically related individuals who share a gene pool

Random mating – selection of a mate without regard to genotype

Recessive – a trait which is expressed only in homozygotes

Recombinant – an individual in a linkage study in whom the marker and disease loci have assorted at parental meiosis

Recombinant DNA – artificial insertion of a portion of DNA from one organism into the genome of another

Recombination – the formation of new combinations of linked genes by crossing over between their loci during meiosis.

Restriction enzyme – an enzyme which cleaves DNA at sequence specific sites (the recognition site)

Restriction fragment length polymorphism (RFLP) – a recognition site for a restriction enzyme which may or may not be present

Reverse transcriptase – an enzyme which can make complementary DNA from messenger RNA

Segregation – the separation of allelic genes at meiosis

Sex-limited – a trait expressed only in one sex

Sex-linked – inheritance of a gene carried on a sex chromosome

Sister chromatid exchange – exchange of DNA by sister chromatids

Species – a set of individuals who can interbreed and have fertile progeny

Spindle fibres – protein fibres connecting centromeres to centrioles at cell division

Sporadic – no known genetic basis

Synteny – loci on the same chromosome which may or may not be linked

Teratogen – any agent which causes congenital malformations

Trait – any gene determined characteristic

Transcription – production of mRNA from the DNA template

Transformation – *in vitro* uptake of plasmids by bacteria

Translation – conversion of the mRNA message to a polypeptide chain

Translocation – the transfer of chromosomal material between chromosomes

Transposable elements – discrete pieces of DNA which are capable of moving from site to site

Triploid – a cell with three times the haploid number of chromosomes

Trisomy – three copies of a given chromosome per cell

Wild type – the normal allele

Zygote – the fertilized ovum

Reference Textbooks

The following textbooks are suggested for further reading and are grouped according to area of interest.

Background general biology

Kimball J.W. (1977) *Biology*, 4th ed. Reading, Mass: Addison-Wesley

Further information about particular genetic diseases

Beighton P. (1978) *Inherited Disorders of the Skeleton*. Edinburgh: Churchill Livingstone.

Bergsma D.S. (1979) *Birth Defects Atlas and Compendium*, 2nd ed. Baltimore: Williams and Wilkins.

Emery A.E.H. & Rimoin D.L. (1983) *The Principles and Practice of Medical Genetics*. Edinburgh: Churchill Livingstone.

McKusick V.A. (1982) *Mendelian Inheritance in Man: Catalogs of Autosomal Dominant, Autosomal Recessive and X-linked Phenotypes*, 5th ed. Baltimore: The Johns Hopkins Press.

Stanbury J.B., Wyngaarden J.B., Fredrickson D.S. & Goldstein J.L. (1983) *The Metabolic Basis of Inherited Disease*, 5th ed. New York, McGraw-Hill.

Warkany J. (1971) *Congenital Malformations*. Chicago: Year Book Medical Publishers.

Syndrome identification

Gorlin R.J., Pindborg J.J. & Cohen M.M. (1976) *Syndromes of the Head and Neck*, 2nd ed. New York: McGraw-Hill.

Smith D.W. (1982) *Recognisable Patterns of Human Malformation*. 3rd ed. Philadelphia: W.B. Saunders.

Genetic counselling

Emery A.E.H. (1976) *Methodology in Medical Genetics*. Edinburgh: Churchill Livingstone.

Harper P.S. (1981) *Practical Genetic Counselling*. Bristol: John Wright.

Index